PATENT TECHNOLOGY:

TRANSFER AND INDUSTRIAL COMPETITION

PATENT TECHNOLOGY:

TRANSFER AND INDUSTRIAL COMPETITION

JUANITA M. BRANES
EDITOR

Nova Science Publishers, Inc.
New York

Copyright © 2007 by Nova Science Publishers, Inc.

All rights reserved. No part of this book may be reproduced, stored in a retrieval system or transmitted in any form or by any means: electronic, electrostatic, magnetic, tape, mechanical photocopying, recording or otherwise without the written permission of the Publisher.

For permission to use material from this book please contact us:
Telephone 631-231-7269; Fax 631-231-8175
Web Site: http://www.novapublishers.com

NOTICE TO THE READER

The Publisher has taken reasonable care in the preparation of this book, but makes no expressed or implied warranty of any kind and assumes no responsibility for any errors or omissions. No liability is assumed for incidental or consequential damages in connection with or arising out of information contained in this book. The Publisher shall not be liable for any special, consequential, or exemplary damages resulting, in whole or in part, from the readers' use of, or reliance upon, this material.

Independent verification should be sought for any data, advice or recommendations contained in this book. In addition, no responsibility is assumed by the publisher for any injury and/or damage to persons or property arising from any methods, products, instructions, ideas or otherwise contained in this publication.

This publication is designed to provide accurate and authoritative information with regard to the subject matter cover herein. It is sold with the clear understanding that the Publisher is not engaged in rendering legal or any other professional services. If legal, medical or any other expert assistance is required, the services of a competent person should be sought. FROM A DECLARATION OF PARTICIPANTS JOINTLY ADOPTED BY A COMMITTEE OF THE AMERICAN BAR ASSOCIATION AND A COMMITTEE OF PUBLISHERS.

Library of Congress Cataloging-in-Publication Data
Available upon request

ISBN 13 978-1-60021-220-8
ISBN 10 1-60021-220-4

Published by Nova Science Publishers, Inc. ✢ New York

CONTENTS

Preface		vii
Chapter 1	Patent Reform: Innovation Issues *Wendy H. Schacht and John R. Thomas*	1
Chapter 2	Patent Law and its Application to the Pharmaceutical Industry: An Examination of the Drug Price Competition and Patent Term Restoration Act of 1984 ("The Hatch-Waxman Act") *Wendy H. Schacht and John R. Thomas*	39
Chapter 3	Industrial Competitiveness and Technological Advancement: Debate over Government Policy *Wendy H. Schacht*	77
Chapter 4	Safe Harbor for Preclinical Use of Patented Inventions in Drug Research and Development: *Merck Kgaa V. Integra Lifesciences I, Ltd.* *Brian T. Yeh*	95
Chapter 5	File-Sharing Software and Copyright Infringement: *Metro-Goldwyn-Mayer Studios, Inc. V. Grokster, Ltd.* *Brian Yeh and Robin Jeweler*	107
Chapter 6	Cooperative R&D: Federal Efforts to Promote Industrial Competitiveness *Wendy H. Schacht*	119
Chapter 7	Technology Transfer: Use of Federally Funded Research and Development *Wendy H. Schacht*	137
Index		155

PREFACE

The pace of U.S. technologies' advancement is crucial to the U.S. economy and its growth. Productivity and international competivieness are important contributing factors. This new book presents important analyses on patents, technology transfer and industrial competitiveness.

Congressional interest in patent policy and possible patent reform has expanded as the importance of intellectual property to innovation has increased. Patent ownership is perceived as an incentive to the technological advancement that leads to economic growth. However, growing interest in patents has been accompanied by persistent concerns about the fairness and effectiveness of the current system. Several recent studies, including those by the National Academy of Sciences and the Federal Trade Commission, have recommended patent reform to address perceived deficiencies in the operation of the patent regime. Other experts maintain that major alterations in existing law are unnecessary and that the patent process can, and is, adapting to technological progress. The Patent Act of 2005, H.R. 2795, introduced in June 2005, would work significant legal reforms to the patent system as reported in chapter 1. Among the more notable of these changes are a shift to a first-inventor-to-file priority system; substantive and procedural modifications to the patent law doctrines of willful infringement and inequitable conduct; and adoption of post-issuance opposition proceedings, prior user rights, and pre-issuance publication of all pending applications. Several of these proposals have been the subject of discussion within the patent community for many years, but others are more novel propositions. Pending legislation attempts to address several issues of concern including the quality of issued patents, the expense and complexity of patent litigation, harmonization of U.S. patent law with the laws of our leading trading partners, potential abuses committed by patent speculators, and the special needs of individual inventors, universities, and small firms with respect to the patent system. In addition, although the existing patent statute in large measure applies the same basic rules to different sorts of inventions, regardless of the technological field of that invention, the patent system is widely believed to impact different industries in varying ways. The provisions of H.R. 2795 would arguably work the most sweeping reforms to the U.S. patent system since the nineteenth century. However, many of these proposals, such as pre-issuance publication, prior user rights, and oppositions, have already been implemented in U.S. law to a more limited extent. These and other reforms, such as the first-inventor-to-file priority system and elimination of the best mode requirement, also reflect the decades-old patent practices of Europe, Japan, and our other leading trading partners. Other knowledgeable observers are nonetheless concerned that certain of these proposals would weaken the patent right, thereby diminishing needed incentives for innovation. Some also believe that changes of this magnitude, occurring at the same time, do not present the most prudent course for the patent system. Patent reform therefore confronts Congress with difficult legal, practical, and policy issues, but also with apparent possibilities for altering and possibly improving the legal regime that has long been recognized as an engine of innovation within the U.S. economy.

Congressional interest in the availability of prescription drugs has focused attention on the role of patents in the pharmaceutical industry. The industry has been described as patent-intensive. Enterprises within this sector frequently obtain patent protection and enforce patent rights, and reportedly place a higher comparative value on patents than do competitors in many other markets. Chapter 2 describes the patent law is based upon the Patent Act of 1952, codified in Title 35 of the United States Code. This statute allows inventors to obtain patents on processes, machines, manufactures, and compositions of matter that are useful, new, and nonobvious. Granted patents confer the right to exclude others from making, using, selling, offering to sell, or importing into the United States the patented invention. The Drug Price Competition and Patent Term Restoration Act of 1984 (the 1984 Act) – commonly known as the "Hatch-Waxman Act" – made several significant changes to the patent laws designed to encourage innovation in the pharmaceutical industry while facilitating the speedy introduction of lower-cost generic drugs. These changes include provisions for extending the term of a patent to reflect regulatory delays encountered in obtaining marketing approval by the Food and Drug Administration (FDA); a statutory exemption from patent infringement for activities associated with regulatory marketing approval; establishment of mechanisms to challenge the validity of a pharmaceutical patent; and a reward for disputing the validity, enforceability, or infringement of a patented and approved drug. The 1984 Act also provides the FDA with certain authorities to offer periods of marketing exclusivity for a pharmaceutical independent of the rights conferred by patents. Many experts agree the 1984 Act has had a significant effect on the availability of generic substitutes for brand name drugs. Lower cost generics tend to be rapidly marketed after patent expiration. Increasing investment in R&D and gains in the research intensity of the pharmaceutical industry appear to indicate that the act has not deterred the development of new drugs. However, some questioned whether the law is needed to achieve the stated goals. Critics maintained the necessity of patent-related incentives for innovation is mitigated by other federal activities. Supporters of the existing approach argued that these incentives are precisely what foster a robust pharmaceutical industry. Of fundamental interest was whether alterations of the act were in order to reflect any perceived changes in the research environment since the legislation was enacted in the 1980s.

There is on-going interest in the pace of U.S. technological advancement due to its influence on U.S. economic growth, productivity, and international competitiveness. Because technology can contribute to economic growth and productivity increases, congressional interest has focused on how to augment private-sector technological development as discussed in chapter 3. Legislative activity over the past two decades has created a policy for technology development, albeit an ad hoc one. Because of the lack of consensus on the scope and direction of a national policy, Congress has taken an incremental approach aimed at creating new mechanisms to facilitate technological advancement in particular areas and making changes and improvements as necessary. Congressional action has mandated specific technology development programs and obligations in federal agencies that did not initially support such efforts. Many programs were created based upon what individual committees judged appropriate within the agencies over which they had authorization or appropriation responsibilities. The use of line item funding for these activities, including the Advanced Technology Program and the Manufacturing Extension Program of the National Institute of Standards and Technology, as well as for the Undersecretary for Technology at the Department of Commerce, is viewed as a way to ensure that the government encourages technological advance in the private sector. Some legislative activity, beginning in the 104th Congress, has been directed at eliminating or significantly curtailing many of these federal efforts. Although this approach has not been successful, the budgets for several programs have declined. Questions have been raised concerning the proper role of the federal government in technology development and the competitiveness of U.S. industry. As the 109th Congress develops its budget priorities, how the government encourages technological progress in the private sector again may be explored and/or redefined.

In *Merck KGaA v. Integra Lifesciences I, Ltd.*, __ U.S. __, 125 S. Ct. 2372 (2005), the United States Supreme Court unanimously held that the preclinical use of patented inventions in drug research is exempted from patent infringement claims by the "safe harbor" provision of the Patent Act, 35 U.S.C. § 271(e)(1). (Merck KGaA is a German company unaffiliated with the U.S.-based pharmaceutical company Merck and Co.) This decision potentially may help expedite the development of new medical treatments and lower the cost of some drugs for consumers which is the focus of chapter 4. In 2003, the U.S. Court of Appeals for the Federal Circuit had narrowly construed the safe harbor provision as protecting only clinical research activities that produce information for submission to the Food and Drug Administration (FDA) in the regulatory process. In vacating that decision, the U.S. Supreme Court ruled that the exemption applies to *all* uses of patented inventions that are "reasonably related" to the process of developing any information for FDA submission. The Court explained that, under certain conditions, the safe harbor provision is even "sufficiently broad" to protect the use of patented compounds in experiments that are not ultimately submitted to the FDA or drug experiments that are not ultimately the subject of an FDA submission. Finally, the scope of the exemption is not limited only to preclinical studies pertaining to a drug's safety in humans, but also includes preclinical data regarding a drug's efficacy, mechanism of action, pharmacokinetics, and pharmacology. However, the Court cautioned that the exemption does not reach all experimental activity that at some point, however attenuated, may lead to an FDA approval process. For example, the safe harbor provision does not embrace basic scientific research performed on a patented compound without the intent to develop a particular drug or without a reasonable belief that the compound will cause a particular physiological effect that the researcher desires. In addition, because the matter was not at issue in the case, the Court expressly declined to decide whether or to what extent the exemption applies to patented "research tools" that are often used to facilitate general research in developing compounds for FDA submissions.

Metro-Goldwyn-Mayer Studios, Inc. v. Grokster, Ltd. is a Ninth Circuit Court of Appeals decision considering allegations of contributory and vicarious copyright infringement by companies which distribute peer-to-peer file-sharing software. The software facilitates direct copyright infringement by its users. It is the first decision to reject infringement claims against and find in favor of companies distributing the software. To date, other digital media file-sharing software decisions have found in favor of the copyright holders, most notably *A and M Records, Inc. v. Napster, Inc.* and *In re: Aimster Copyright Litigation*. But in *Grokster*, the court granted summary judgment for the software companies. This report provides a general overview of peer-to-peer file-sharing technology and then examines why *Grokster* produced a result different from those in other peer-to-peer software litigation. As reported in chapter 5, on December 10, 2004, the U.S. Supreme Court granted a writ of certiorari to hear an appeal in the *Grokster* case. One explanation for *Grokster* is the differences in the technological design of the various peer-to-peer systems. While the pioneering file-sharing network Napster provided exclusively for the exchange of audio files, the software companies sued in *Grokster* employ more advanced peer-to-peer technology that allows the additional sharing of video clips, text documents, and computer programs. The *Grokster* court acknowledged these expanded capabilities as legitimate uses of the software, and thus became the first court to accept the "substantial, noninfringing uses" defense to copyright infringement liability, a defense developed by the U.S. Supreme Court in *Sony Corp. of America v. Universal City Studios, Inc.* Two months after the U.S. district court decision in *Grokster*, the Seventh Circuit Court of Appeals in *Aimster* also expressed qualified support for application of the *Sony* defense to file-sharing software, but nevertheless upheld a preliminary injunction against Aimster because the software company failed to demonstrate that its peer-to-peer service was ever actually used for any substantial noninfringing purposes. Another factor determinative to the *Grokster* court was the software companies' limited contribution to the infringing activity of its software users, and their limited ability to police their networks. Whereas Napster actively provided on-going services and technical support to its users in locating and downloading music files, the *Grokster* defendants distribute software that operates across peer-to-peer networks outside of their control and supervision. This more sophisticated software allows users to connect to each other and swap files directly, without the need for a centralized search index or website to facilitate the file transfers, as Napster had maintained.

As reported in chapter 6, in response to the foreign challenge in the global marketplace, the United States Congress has explored ways to stimulate technological advancement in the private sector. The government has supported various efforts to promote cooperative research and development activities among industry, universities, and the federal R&D establishment designed to increase the competitiveness of American industry and to encourage the generation of new products, processes, and services. Among the issues before Congress are whether joint ventures contribute to industrial competitiveness and what role, if any, the government has in facilitating such arrangements. Collaborative ventures are intended to accommodate the strengths and responsibilities of all sectors involved in innovation and technology development. Academia, industry, and government often have complementary functions. Joint projects allow for the sharing of costs, risks, facilities, and expertise. Cooperative activity covers various institutional and legal arrangements including industry-industry, industry-university, and industry-government efforts. Proponents of joint ventures argue that they permit work to be done that is too expensive for one company to support and allow for R&D that crosses traditional boundaries of expertise and experience. Such arrangements make use of existing, and support the development of new, resources, facilities, knowledge, and skills. Opponents argue that these endeavors dampen competition necessary for innovation. Federal efforts to encourage cooperative activities include the National Cooperative Research Act; the National Cooperative Production Act; tax changes permitting credits for industry payments to universities for R&D and deductions for contributions of equipment used in academic research; and amendments to the patent laws vesting title to inventions made under federal funding in universities. Technology transfer from the government to the private sector is facilitated by several laws. In addition, there are various ongoing cooperative programs supported by various federal departments and agencies. Given the increased popularity of cooperative programs, questions might be raised as to whether they are meeting expectations. It may be too soon to determine the effectiveness of the joint R&D venture as a mechanism to increase technological advancement in the United States. There is often a long time lag between research and the availability of a product, process, or service. Many of the collaborative activities fostered by the federal government are of recent origin and therefore have not had sufficient time to generate measurable results. However, raising certain issues might serve to develop a framework for addressing future, near-term decisions concerning technology development and cooperative R&D. These include questions about the emphasis on collaborative ventures in research rather than in technology development; cooperative manufacturing; defense vs. civilian support; and access by foreign companies.

The government spends approximately one third of the $83 billion federal R&D budget for intramural research and development to meet mission requirements in over 700 government laboratories (including Federally Funded Research and Development Centers). The technology and expertise generated by this endeavor may have application beyond the immediate goals or intent of federally funded R&D. This can be achieved by technology transfer, a process by which technology developed in one organization, in one area, or for one purpose is applied in another organization, in another area, or for another purpose. It is a way for the results of the federal R&D enterprise to be used to meet other national needs, including the economic growth that flows from new commercialization in the private sector; the government's requirements for products and processes to operate effectively and efficiently; and the demand for increased goods and services at the state and local level. Chapter 6 discusses the system Congress has established to facilitate the transfer of technology to the private sector and to state and local governments. Despite this, use of federal R&D results has remained restrained, although there has been a significant increase in private sector interest and activities over the past several years. Critics argue that working with the agencies and laboratories continues to be difficult and time-consuming. Proponents of the current effort assert that while the laboratories are open to interested parties, the industrial community is making little effort to use them. At the same time, State governments are increasingly involved in the process. At issue is whether incentives for technology transfer remain necessary, if additional legislative initiatives are needed to encourage increased technology transfer, or if the responsibility to use the available resources now rests with the private sector.

In: Patent Technology
Editor: Juanita M. Branes, pp. 1-37

ISBN: 978-1-60021-220-8
© 2007 Nova Science Publishers, Inc.

Chapter 1

PATENT REFORM: INNOVATION ISSUES[*]

Wendy H. Schacht and John R. Thomas

ABSTRACT

Congressional interest in patent policy and possible patent reform has expanded as the importance of intellectual property to innovation has increased. Patent ownership is perceived as an incentive to the technological advancement that leads to economic growth. However, growing interest in patents has been accompanied by persistent concerns about the fairness and effectiveness of the current system. Several recent studies, including those by the National Academy of Sciences and the Federal Trade Commission, have recommended patent reform to address perceived deficiencies in the operation of the patent regime. Other experts maintain that major alterations in existing law are unnecessary and that the patent process can, and is, adapting to technological progress.

The Patent Act of 2005, H.R. 2795, introduced in June 2005, would work significant legal reforms to the patent system. Among the more notable of these changes are a shift to a first-inventor-to-file priority system; substantive and procedural modifications to the patent law doctrines of willful infringement and inequitable conduct; and adoption of post-issuance opposition proceedings, prior user rights, and pre-issuance publication of all pending applications. Several of these proposals have been the subject of discussion within the patent community for many years, but others are more novel propositions.

Pending legislation attempts to address several issues of concern including the quality of issued patents, the expense and complexity of patent litigation, harmonization of U.S. patent law with the laws of our leading trading partners, potential abuses committed by patent speculators, and the special needs of individual inventors, universities, and small firms with respect to the patent system. In addition, although the existing patent statute in large measure applies the same basic rules to different sorts of inventions, regardless of the technological field of that invention, the patent system is widely believed to impact different industries in varying ways.

The provisions of H.R. 2795 would arguably work the most sweeping reforms to the U.S. patent system since the nineteenth century. However, many of these proposals, such as pre-issuance publication, prior user rights, and oppositions, have already been

---

[*] Excerpted from CRS Report RL32996, dated July 15, 2005.

implemented in U.S. law to a more limited extent. These and other reforms, such as the first-inventor-to-file priority system and elimination of the best mode requirement, also reflect the decades-old patent practices of Europe, Japan, and our other leading trading partners.

Other knowledgeable observers are nonetheless concerned that certain of these proposals would weaken the patent right, thereby diminishing needed incentives for innovation. Some also believe that changes of this magnitude, occurring at the same time, do not present the most prudent course for the patent system. Patent reform therefore confronts Congress with difficult legal, practical, and policy issues, but also with apparent possibilities for altering and possibly improving the legal regime that has long been recognized as an engine of innovation within the U.S. economy.

INTRODUCTION

Congressional interest in patent reform has increased as the patent system becomes more significant to U.S. industry. There is broad agreement that more patents are sought and enforced then ever before; that the attention paid to patents in business transactions and corporate boardrooms has dramatically increased; and that the commercial and social significance of patent grants, licenses, judgments, and settlements is at an all-time high.[1] As the United States becomes even more of a high-technology, knowledge-based economy, the importance of patents may grow even further in the future.

Increasing interest in patents has been accompanied by persistent concerns about the fairness and effectiveness of the current system. Several recent studies, including those by the National Academy of Sciences and the Federal Trade Commission, have recommended patent reform to address perceived deficiencies in the operation of the patent regime.[2] Other experts maintain that major alterations in existing law are unnecessary and that the patent process can, and is, adapting to technological progress.

In June 2005, legislation was introduced that attempts to respond to current concerns about the functioning of the patent process. The Patent Act of 2005, H.R. 2795, proposes significant legal reforms to the patent system, including a shift to a first-inventor-to-file priority system; substantive and procedural modifications to the patent law doctrines of willful infringement and inequitable conduct; and adoption of post-issuance opposition proceedings, prior user rights, and pre-issuance publication of all pending applications. Several of these proposals have been the subject of discussion within the patent community for many years, but others are more novel propositions.

This report provides an overview of current patent reform issues. It begins by offering a summary of the structure of the current patent system and the role of patents in innovation policy. The report then reviews some of the broader issues and concerns, including patent quality, the high costs of patent litigation, international harmonization, and speculation in patents, that have motivated H.R. 2795's diverse reform proposals. The specific components of this legislation are then identified and reviewed in greater detail.

PATENTS AND INNOVATION POLICY

The Mechanics of the Patent System

The patent system is grounded in Article I, Section 8, Clause 8 of the U.S. Constitution, which states that "The Congress Shall Have Power . . . To promote the Progress of Science and useful Arts, by securing for limited Times to Authors and Inventors the exclusive Right to their respective Writings and Discoveries . . . ." As mandated by the Patent Act of 1952,[3] U.S. patent rights do not arise automatically. Inventors must prepare and submit applications to the U.S. Patent and Trademark Office (USPTO) if they wish to obtain patent protection.[4] USPTO officials known as examiners then assess whether the application merits the award of a patent.[5] The patent acquisition process is commonly known as "prosecution."[6]

In deciding whether to approve a patent application, a USPTO examiner will consider whether the submitted application fully discloses and distinctly claims the invention.[7] In addition, the application must disclose the "best mode," or preferred way, that the applicant knows to practice the invention.[8] The examiner will also determine whether the invention itself fulfills certain substantive standards set by the patent statute. To be patentable, an invention must be useful, novel and nonobvious. The requirement of usefulness, or utility, is satisfied if the invention is operable and provides a tangible benefit.[9] To be judged novel, the invention must not be fully anticipated by a prior patent, publication or other knowledge within the public domain.[10] A nonobvious invention must not have been readily within the ordinary skills of a competent artisan at the time the invention was made.[11]

If the USPTO allows the patent to issue, the patent proprietor obtains the right to exclude others from making, using, selling, offering to sell or importing into the United States the patented invention.[12] Those who engage in these acts without the permission of the patentee during the term of the patent can be held liable for infringement. Adjudicated infringers may be enjoined from further infringing acts.[13] The patent statute also provides for the award of damages "adequate to compensate for the infringement, but in no event less than a reasonable royalty for the use made of the invention by the infringer."[14]

The maximum term of patent protection is ordinarily set at 20 years from the date the application is filed.[15] At the end of that period, others may employ that invention without regard to the expired patent.

Patent rights are not self-enforcing. Patentees who wish to compel others to observe their rights must commence enforcement proceedings, which most commonly consist of litigation in the federal courts. Although issued patents enjoy a presumption of validity, accused infringers may assert that a patent is invalid or unenforceable on a number of grounds.[16] The U.S. Court of Appeals for the Federal Circuit (Federal Circuit) possesses national jurisdiction over most patent appeals from the district courts.[17] The U.S. Supreme Court enjoys discretionary authority to review cases decided by the Court of Appeals for the Federal Circuit.[18]

Innovation Policy

Patent ownership is perceived to be an incentive to innovation, the basis for the technological advancement that contributes to economic growth. It is through the commercialization and use of new products and processes that productivity gains are made and the scope and quality of goods and services are expanded. Award of a patent is intended to stimulate the investment necessary to develop an idea and bring it to the marketplace embodied in a product or process. Patent title provides the recipient with a limited-time monopoly over the use of his discovery in exchange for the public dissemination of information contained in the patent application. This is intended to permit the inventor to receive a return on the expenditure of resources leading to the discovery but does not guarantee that the patent will generate commercial benefits. The requirement for publication of the patent is expected to stimulate additional innovation and other creative means to meet similar and expanded demands in the marketplace.

Innovation produces new knowledge. One characteristic of this knowledge is that it is a "public good," a good that is not consumed when it is used. This "public good" concept underlies the U.S. patent system. Absent a patent system, "free riders" could easily duplicate and exploit the inventions of others. Further, because they incurred no cost to develop and perfect the technology involved, copyists could undersell the original inventor. The resulting inability of inventors to capitalize on their inventions would lead to an environment where too few inventions are made.[19] The patent system corrects this market failure problem by providing innovators with an exclusive interest in their inventions, thereby allowing them to capture its marketplace value.

The regime of patents purportedly serves other goals as well. The patent system encourages the disclosure of products and processes, for each issued patent must include a description sufficient to enable skilled artisans to practice the patented invention.[20] At the close of the patent's twenty-year term,[21] others may practice the claimed invention without regard to the expired patent. In this manner the patent system ultimately contributes to the growth of the public domain.

Even during their term, issued patents may also encourage others to "invent around" the patentee's proprietary interest. A patentee may point the way to new products, markets, economies of production and even entire industries. Others can build upon the disclosure of a patent instrument to produce their own technologies that fall outside the exclusive rights associated with the patent.[22]

The patent system has also been identified as a facilitator of markets. Absent patent rights, an inventor may have scant tangible assets to sell or license. In addition, an inventor might otherwise be unable to police the conduct of a contracting party. Any technology or know-how that has been disclosed to a prospective licensee might be appropriated without compensation to the inventor. The availability of patent protection decreases the ability of contracting parties to engage in opportunistic behavior. By lowering such transaction costs, the patent system may make technology-based transactions more feasible.[23]

Through these mechanisms, the patent system can act in more socially desirable ways than its chief legal alternative, trade secret protection. Trade secrecy guards against the improper appropriation of valuable, commercially useful and secret information. In contrast to patenting, trade secret protection does not result in the disclosure of publicly valuable information. That is because an enterprise must take reasonable measures to keep secret the information for which trade secret protection is sought. Taking the steps necessary to maintain secrecy, such as implementing physical security measures, also imposes costs that may ultimately be unproductive for society.[24]

The patent system has long been subject to criticism, however. Some observers have asserted that the patent system is unnecessary due to market forces that already suffice to create an optimal level of innovation. The desire to obtain a lead time advantage over competitors, as well as the recognition that technologically backward firms lose out to their rivals, may well provide sufficient inducement to invent without the need for further incentives.[25] Other commentators believe that the patent system encourages industry concentration and presents a barrier to entry in some markets.[26] Still other observers believe that the patent system too frequently attracts speculators who prefer to acquire and enforce patents rather than engage in socially productive activity.[27]

When analyzing the validity of these competing views, it is important to note the lack of rigorous analytical methods available for studying the effect of the patent law upon the U.S. economy as a whole. The relationship between innovation and patent rights remains poorly understood. As a result, current economic and policy tools do not allow us to calibrate the patent system precisely in order to produce an optimal level of investment in innovation. Thus, each of the arguments for and against the patent system remains open to challenge by those who are unpersuaded by their internal logic.

CURRENT ISSUES AND CONCERNS

Pending legislation proposes a number of changes to diverse aspects of the patent system. Although these reforms are undoubtedly motivated by a range of concerns, a discrete number of issues have been the subject of persistent discussion in the patent community over a period of many years. Among these issues are concern for the quality of issued patents, the expense and complexity of patent litigation, harmonization of U.S. patent law with the laws of our leading trading partners, potential abuses committed by patent speculators, and the special needs of individual inventors, universities, and small firms with respect to the patent system. In addition, although the patent statute in large measure applies the same basic rules to different sorts of inventions, regardless of the technological field of that invention, the patent system is widely believed to impact different industries in varying ways.[28] As a result, different industries can be expected to espouse different views of certain patent reform proposals. Before turning to a more specific analysis of individual legislative proposals, this report reviews the proposed legislation's broader themes with regard to these issues and concerns.

Patent Quality

Government, industry, academia and the patent bar alike have long insisted that the USPTO approve only those patent applications that describe and claim a patentable advance.[29] Because they meet all the requirements imposed by the patent laws, quality patents may be dependably enforced in court and employed as a technology transfer tool.

Such patents are said to confirm private rights by making their proprietary uses, and therefore their value, more predictable. Quality patents also may clarify the extent that others may approach the protected invention without infringing. These traits in turn should strengthen the incentives of private actors to engage in value-maximizing activities such as innovation or commercial transactions.[30]

In contrast, poor patent quality is said to hold deleterious consequences. Large numbers of inappropriately granted patents may negatively impact entrepreneurs. For example, innovative firms may be approached by an individual with a low quality patent that appears to cover the product they are marketing. The innovative firm may recognize that the cost of challenging a patent even of dubious validity may be considerable. Therefore, the firm may choose to make payments under licensing arrangements, or perhaps decide not to market its product at all, rather than contest the patent proprietor's claims.[31]

Poor patent quality may also encourage opportunistic behavior. Perhaps attracted by large damages awards and a potentially porous USPTO, rent-seeking entrepreneurs may be attracted to form speculative patent acquisition and enforcement ventures. Industry participants may also be forced to expend considerable sums on patent acquisition and enforcement.[32] The net results would be reduced rates of innovation, decreased patent-based transactions, and higher prices for goods and services.

Although low patent quality appears to affect both investors and competitors of a patentee, patent proprietors themselves may also be negatively impacted. Patent owners may make managerial decisions, such as whether to build production facilities or sell a product, based upon their expectation of exclusive rights in a particular invention. If their patent is declared invalid by the USPTO or a court, the patentee will be stripped of exclusive rights without compensation. The issuance of large numbers of invalid patents would increase the possibility that the investment-backed expectations of patentees would be disappointed.[33]

The notion that high patent quality is socially desirable has been challenged, however. Some commentators believe that market forces will efficiently assign patent rights no matter what their quality. Others observe that few issued patents are the subject of litigation and further estimate that only a minority of patents are licensed or sold. Because many patented inventions are not used in a way that calls their validity into question, some observers maintain, society may be better off making a detailed review into the patentability of an invention only in those few cases where that invention is of commercial significance.[34]

Proposed legislation bears upon the patent quality issue. In particular, H.R. 2795 would allow for increased public participation in USPTO decisionmaking through a pre-issuance submission procedure. This bill would also allow for post-issuance opposition proceedings, which would potentially allow interested parties to "weed out" invalid patents before they are the subject of licensing or infringement litigation.

Litigation Costs

Patent enforcement is often expensive. The complex legal and technological issues, extensive discovery proceedings, expert witnesses, and specially qualified attorneys associated with patent trials can lead to high costs.[35] One study published in 2000 concluded that the average cost of patent enforcement was $1.2 million.[36] These expenses appear to be increasing, with one more recent commentator describing an "industry rule of thumb" whereby "any patent infringement lawsuit will easily cost $1.5 million in legal fees alone to defend."[37] Higher stakes litigation is even more costly: For patent suits involving damages claims of more than $25 million, expenses reportedly increase to $4 million per side.[38]

For innovative firms that are not infrequently charged with patent infringement, or that bring claims of patent infringement themselves, the annual expenses associated with patent litigation can be very dear. The Microsoft Corporation reportedly defends an average of 35 to 40 patent lawsuits annually at a cost of almost $100 million.[39] The Intel Corporation has recently been estimated to spend $20 million a year on patent litigation.[40]

The high costs of litigation may discourage patent proprietors from bringing meritorious claims against infringers. They may also encourage firms to license patents of dubious merit rather than contest them in court. Pending legislation endeavors to make patent litigation less costly and complex through modification of the patent law doctrines of willful infringement and inequitable conduct. It would also call for an administrative opposition proceeding that, in some measure, could serve as a less expensive alternative to litigation.

International Harmonization

In our increasingly globalized, high-technology economy, patent protection in a single jurisdiction is often ineffective to protect the interests of inventors. As a result, U.S. inventors commonly seek patent protection abroad. Doing so can be a costly, time-consuming, and difficult process. There is no global patent system. Inventors who desire intellectual property protection in a particular country must therefore take specific steps to procure a patent within that jurisdiction.[41]

Differences in national laws are among the difficulties faced by U.S. inventors seeking patent rights overseas. Although the world's patent laws have undergone considerable harmonization in recent years, several notable distinctions between U.S. patent law and those of our leading trading partners persist. Pending legislation would address some of these differences by modifying U.S. patent law in order to comply with international standards. Among these proposed reforms are adoption of a first-inventor-to-file priority system, a post-issuance opposition system, assignee filing, publication of all pending patent applications, and prior user rights; elimination of the best mode requirement; and encouragement for the adoption of a one-year grace period within the European Patent Convention and Japanese Patent Act.

Potential Abuses of Patent Speculators

Some commentators believe that the patent system too frequently attracts speculators who prefer to acquire and enforce patents rather than engage in research, development, manufacturing, or other socially productive activity.[42] Patent speculators are sometimes termed "trolls," after creatures from folklore that would emerge from under a bridge in order to waylay travelers.[43] The late Jerome C. Lemelson, a prolific inventor who owned hundreds of patents and launched numerous charges of patent infringement, has sometimes been mentioned in this context. The annual revenue of the Lemelson estate's patent licensing program has been reported as in excess of $1.5 billion.[44] But as explained by journalist Michael Ravnitsky, "critics charge that many Lemelson patents are so-called submarine patents, overly broad applications that took so long to issue or were so general in nature that their owners could unfairly claim broad infringement across entire industry sectors."[45] Of such patent ventures, patent attorney James Pooley observes:

> Of course there is nothing inherently wrong with charging someone rent to use your property, including intellectual property like patents. But it's useful to keep in mind — especially when listening to prattle about losing American jobs to foreign competition — that these patent mills produce no products. Their only output is paper, of a highly threatening sort.[46]

Patent enforcement suits brought by patent speculators appear to present special concerns for manufacturers and service providers. If one manufacturer or service provider commences litigation against another, the defendant can often counter with its own claims of patent infringement against the plaintiff. Because patent speculators do not otherwise participate in the marketplace, however, they are immune to such counterclaims. This asymmetry in litigation positions reportedly reduces the bargaining power of manufacturers and service providers and exposes them to harassment.[47]

Observers hasten to note, however, that not every patent proprietor who does not commercialize the patented invention should properly be considered an opportunistic "troll." A nonmanufacturing patentee may lack the expertise or resources to produce a patented product, prefer to commit itself to further innovation, or otherwise have legitimate reasons for its behavior.[48] Universities and small biotechnology companies often fit into this category. Further, whether classified as a "troll" or not, each patent owner has presumptively fulfilled all of the relevant statutory requirements. Among these obligations is a thorough disclosure of a novel, nonobviousn invention to the public.[49]

Pending legislation would impact concerns over "trolling" by amendment to the patent statute's provision regarding injunctions, as well as allowing limitations upon so-called continuation practice at the USPTO.

The Role of Individuals, Universities and Small Entities

Entrepreneurs and small, innovative firms play a role in the technological advancement and economic growth of the United States.[50] Several studies commissioned by U.S. federal agencies have concluded that individuals and small entities constitute a significant source of innovative products and services.[51] Studies have also indicated that entrepreneurs and small, innovative firms rely more heavily upon the patent system than larger enterprises. Larger companies are said to possess alternative means for achieving a proprietary or property-like interest in a particular technology. For example, trade secrecy, ready access to markets, trademark rights, speed of development, and consumer goodwill may to some degree act as substitutes to the patent system.[52] However, individual inventors and small firms often do not have these mechanisms at their disposal. As a result, the patent system may enjoy heightened importance with respect to these enterprises.[53]

In recent years, universities have also become more full-fledged participants in the patent system. This trend has been attributed to the Bayh-Dole Act,[54] a federal statute that allowed universities and other government contractors to retain patent title to inventions developed with the benefit of federal funding.[55] In recent years there has reportedly "been a dramatic increase in academic institutions' investments in technology licensing activities."[56] This increase has been reflected in the growth in the number of patents held by universities, the number of universities with technology transfer offices, and the amount of patent-based licensing revenues that these offices have raised.[57]

The U.S. patent system have long acknowledged the role, and particular needs, of independent inventors, small firms, and universities. For example, the patent statute calls for each of these entities to receive a 50% discount on many USPTO fees.[58] As the USPTO is currently entirely funded by the fees it charges its users,[59] this provision effectively calls for larger institutions to subsidize the patent expenditures of their smaller competitors.

Beyond potentially diminished financial resources vis-a-vis larger concerns, however, observers have disagreed over whether independent inventors, small firms, and universities have particular needs with respect to the patent system, and if so whether those needs should be reflected in patent law doctrines. With respect to the proposed system of "prior user rights,"[60] for example, some observers state that such rights would particularly benefit small entities, which may often lack a sophisticated knowledge of the patent system.[61] Others disagree, stating that smaller concerns rely heavily on the exclusivity of the patent right, and that the adoption of prior user rights would advantage large enterprises.[62] Similar debates have occurred with respect to other patent reform proposals, perhaps reflecting the fact that the community of independent inventors, small firms, and universities is itself a diverse one.

Provisions of pending legislation that appear to be of particular interest to independent inventors, universities, and small businesses include a shift to a first-inventor-to-file priority system, prior user rights, pre-issuance publication of all pending patent applications, and post-issuance oppositions.

Different Roles for Patents in Distinct Industries

To a large extent, the patent statute subjects all inventions to the same standards, regardless of the field in which those inventions arose. Whether the invention is an automobile engine, semiconductor, or a pharmaceutical, it is for the most part subject to the same patentability requirements, scope of rights, and term of protection. Both experience and economic research suggest that distinct industries experience the patent system in different ways, however.[63] As a result, it can be expected that particular industries will react differently to the various patent reform proposals currently before Congress.

Although broad generalizations should be drawn with care, two industries widely perceived as viewing the patent system in different ways are the pharmaceutical and software sectors. Within the pharmaceutical industry, individual patents are perceived as critical to a business model that provides life-saving and life-enhancing medical innovations, but eventually allows members of the public access to medicines at low cost. In particular, often only a handful, and sometimes only one or two patents cover a particular drug product. Patents are also judged to be crucial to the pharmaceutical sector because of the relative ease of replicating the finished product. For example, while it is expensive, complicated, and time consuming to duplicate an airplane, it is relatively simple to perform a chemical analysis of a pill and reproduce it.[64]

In contrast to the pharmaceutical field, the nature of software development is such that innovations are typically cumulative and new products often embody numerous patentable inventions. This environment has led to what has been described as a

> poor match between patents and products in the [software] industry: it is difficult to patent an entire product in the software industry because any particular product is likely to include dozens if not hundreds of separate technological ideas.[65]

This situation may be augmented by the multiplicity of patents often associated with a finished computer product that utilizes the software. It is not uncommon for thousands of different patents (relating to hardware and software) to be embodied in one single computer. In addition, ownership of these patents may well be fractured among hundreds or thousands of different individuals and firms.

In summary, then, the patent laws provide a "one size fits all" system, where all inventions are subject to the same requirements of patentability and scope of protection, regardless of the technical field in which they arose. Innovators in different fields nonetheless have varying experiences with the patent system. These discrepancies, among others, lead to the expectation that distinct industries may react differently to the various patent reform proposals presently considered by Congress.

PROPOSED LEGISLATIVE INITIATIVES

Pending legislation proposes a diverse array of patent reforms. The remainder of this report identifies and reviews these proposals.

First Inventor to File

Pending legislation would alter the U.S. patent priority rule from the current "first-to-invent" principle to the "first-inventor-to-file" principle.[66] Within the patent law, the priority rule addresses the circumstance where two or more persons independently develop the identical or similar invention at approximately the same time. In such cases the patent law must establish a rule as to which of these inventors obtains entitlement to a patent.[67]

In the United States, when more than one patent application is filed claiming the same invention, the patent will be awarded to the applicant who was the first inventor in fact. This conclusion holds even if the first inventor was not the first person to file a patent application directed towards that invention.[68] Under this "first-to-invent" system,[69] the timing of real-world events, such as the date a chemist conceived of a new compound or a machinist constructed a new engine, is of significance.

In every patent-issuing nation except the United States, priority of invention is established by the earliest effective filing date of a patent application disclosing the claiming invention.[70] Stated differently, the inventor who first filed an application at the patent office is presumptively entitled to the patent. Whether or not the first applicant was actually the first individual to complete the invention in the field is irrelevant. This priority system follows the "first-inventor-to file" principle.

A simple example illustrates the distinction between these priority rules. Suppose that inventor A synthesizes a new chemical compound on August 1, 2005, and files a patent application on November 1, 2005 claiming that compound. Suppose further that inventor B independently invents the same compound on September 1, 2005, and files a patent application on October 1, 2005. Inventor A would be awarded the patent under the first-to-invent rule, while Inventor B would obtain the patent under the first-inventor-to-file principle.

Under the current U.S. first-to-invent rule, the majority of priority disputes in the United States are resolved via "interference" proceedings conducted at the USPTO.[71] An interference is a complex administrative proceeding that ordinarily results in the award of priority to one of its participants. These proceedings are not especially common. One estimate is that less than one-quarter of one percent of patents are subject to an interference.[72] This statistic may mislead, however, because the expense of interference cases may result in their use only for the most commercially significant inventions.

The patent community has witnessed an extensive and sometimes emotional debate on the relative merits of the first-to-invent and first-inventor-to-file principle. Supporters of the current first-to-invent principle in part assert that the first-inventor-to-file system would create inequities by sponsoring a "race to the Patent Office." They are also concerned that the first-to-file system would encourage premature and sketchy technological disclosures in hastily-filed patent applications.[73]

Supporters of the first-inventor-to-file principle in part assert that it provides a definite, readily determined and fixed date of priority of invention, which would lead to greater legal certainty within innovative industries. They also contend that the first-inventor-to-file principle would decrease the complexity, length and expense associated with current USPTO interference proceedings. Rather than being caught up in lengthy interference proceedings in an attempt to prove dates of inventive activity that occurred many years previously, they assert, inventors could continue to go about the process of innovation. Supporters also observe that informed U.S. firms already organize their affairs on a first-inventor-to-file basis in order to avoid forfeiture of patent rights abroad.[74]

The effect of a shift to the first-inventor-to-file rule upon individual inventors, small firms, and universities has been debated. Some observers state that such entities often possess fewer resources and wherewithal than their larger competitors, and thus are less able to prepare and file patent applications quickly. Others disagree, stating that smaller concerns are more nimble than larger ones and thus better able to submit applications promptly. They also point to the availability of provisional applications,[75] asserting that such applications allow small entities to secure priority rights readily without a significant expenditure of resources. A quantitative study of interference proceedings by Gerald Mossinghoff, a former Commissioner of the USPTO, also suggested that the first-to-invent rule neither advantaged nor disadvantaged small entities vis-a-vis larger enterprises.[76]

The role of the U.S. Constitution is sometimes debated within the context of the patent priority principle. Article I, section 8, clause 8 of the Constitution provides Congress with the authority to award "inventors" with exclusive rights. Some observers suggest this language suggests, or possibly even mandates, the current first-to-invent system. Others conclude that because the first-inventor-to-file only awards patents to individuals who actually developed the invention themselves, rather than derived it from another, this priority system is permissible under the Constitution.[77]

In weighing the validity of this position, it should be noted that under well-established U.S. law, the first-inventor-in-fact does not always obtain entitlement to a patent. If, for example, a first-inventor-in-fact maintained his invention as a trade secret for many years before seeking patent protection, he may be judged to have "abandoned, suppressed or concealed" the invention.[78] In such a case a second-inventor-in-fact may be awarded a patent on that invention. Courts have reasoned that this statutory rule encourages individuals to disclose their inventions to the public promptly, or give way to an inventor who in fact does so.[79] As the first-inventor-to-file rule acts in a similar fashion to this longstanding patent law principle, conflict between this rule and the Constitution appears unlikely.

Notably, a first-inventor-to-file priority rule does not permit one individual to copy another's invention and then, by virtue of being the first to file a patent application, be entitled to a patent. All patent applicants must have originated the invention themselves, rather than derived it from another.[80] In order to police this requirement, H.R. 2795 provides for "inventor's rights contests" that would allow the USPTO to determine which applicant is entitled to a patent on a particular invention.[81]

Grace Period

Pending legislation would also impact the existing one-year "grace period" enjoyed by U.S. inventors. Current U.S. patent law essentially provides inventors with a one-year period to decide whether patent protection is desirable, and, if so, to prepare an application. Specified activities that occur before the "critical date" —patent parlance for the day one year before the application was filed — will prevent a patent from issuing.[82] If, for example, an entrepreneur first discloses an invention by publishing an article in a scientific journal, she knows that she has one year from the publication date in which to file a patent application. Importantly, uses, sales, and other technical disclosures by third parties will also start the one-year clock running. As a result, inventors have a broader range of concerns than merely their own behavior.[83]

In contrast, many other patent-granting states provide more limited grace periods, or no grace periods at all. In Europe, any sales or publication of an invention anywhere in the world prior to the filing date defeats the patentability of an invention.[84] The Japanese patent system includes a six-month grace period tied only to the activities of the inventor.[85] Under the patent law of Japan, any disclosures of an invention made by a third party even one day before the filing date will prevent the issuance of a patent.

The proposed legislation would make two changes to U.S. patent law. First, the one-year grace period would apply only to the inventor's own activities. Third party activity that occurred even one day before the filing date would constitute prior art and potentially be patent-defeating. This rule represents a change from current law, under which a patent applicant may possibly "antedate" prior art by showing that she invented the subject matter of the application prior to the date of the third party reference.[86] This change is consistent with the proposed shift to a first-inventor-to-file priority rule from a first-to-invent system.

Second, the proposed legislation would include a provision that might encourage adoption of a one-year grace period for inventor activities in Europe and Japan. An understanding of this proposal requires some background information on the international priority system established by the Paris Convention.[87] The international priority system allows an inventor to file a patent application in one Paris Convention signatory state, which is usually the inventor's home country. If the inventor subsequently files patent applications in any other Paris Convention signatory state within the next 12 months, overseas patent-granting authorities will treat the application as if it were filed on the first filing date. Critically, information that enters the public domain between the priority date and subsequent filing dates does not prejudice the later applications. Paris Convention priority allows U.S. inventors to preserve their original USPTO filing dates as they make arrangements to file patent applications overseas.[88]

Suppose, for example, that an inventor files a patent application at the USPTO on October 1, 2005. The inventor then files a patent application claiming the same invention in the Japanese Patent Office on September 1, 2006. As part of his Japanese application, the inventor informs the Japanese Patent Office of the earlier U.S. application. Because Japan has acceded to the Paris Convention, the Japanese Patent Office will treat that inventor's application as if it had been filed on October 1, 2005. As a result, information that entered the public domain after the U.S. filing date would not prejudice the inventor's Japanese application. A journal article published on January 1, 2006, for example, would not limit the opportunity of the inventor to obtain a Japanese patent.

The U.S. patent statute currently limits the usefulness of the Paris Convention priority date for foreign inventors seeking U.S. patent rights. Section 119 of the Patent Act states that:

> no patent shall be granted on any application for patent for an invention which had been patented or described in a printed publication in any country more than one year before the date of the actual filing of the application in this country, or which had been in public use or on sale in this country more than one year prior to such filing.[89]

The effect of this language is that the one-year grace period is measured not from the Paris Convention international priority date, but the actual U.S. filing date.

This limitation may discourage U.S. trading partners from adopting a grace period analogous to that of U.S. law. Consider, for example, a Japanese inventor who publishes an article in a scientific journal describing his new invention on August 1, 2004. Consistent with Japanese patent law, he then files a patent application at the Japanese Patent Office six months later, on February 1, 2005. Then, in accordance with the Paris Convention, he files an application at the USPTO on February 1, 2006.

Under these circumstances, the U.S. patent application should be denied, even though the Japanese inventor appeared to comply with all legal formalities. Because the U.S. patent statute compels the USPTO to assess the grace period as ending as the actual U.S. filing date in 2006, rather than the Paris Convention priority date in 2005, the U.S. patent is barred from issuance. This state of affairs may give pause to nations considering adopting a U.S.-style grace period. Foreign applicants who rely upon grace periods within their own national systems may be put in a position of forfeiture of their U.S. patent rights.

Apparently aware of this concern, pending legislation potentially changes the date the grace period closes from the actual U.S. filing date to the Paris Convention priority date — provided that Europe and Japan adopt laws analogous to that of proposed U.S. law. In the language of H.R. 2795:

Before the date, if ever, that the Director of the [USPTO] publishes a notice . . . declaring that both the European Patent Convention and the patent laws of Japan afford inventors seeking patents a 1-year period prior to the effective filing date of a claimed invention during which disclosures made by the inventor or by others who obtained the subject matter disclosed directly or indirectly from the inventor do not constitute prior art, the term "effective filing date" as used in section 102(a)(1)(A) of title 35, United States Code, shall be construed by disregarding any right of priority except that provided under section 119(e) of title 35, United States Code.[90]

Should the USPTO Director publish a notice in keeping with this provision, then foreign inventors would be able to rely upon their domestic grace periods and maintain their ability to obtain patents in the United States.

Elimination of Sections 102(c), (d) and (f)

Pending legislation would eliminate three provisions of the Patent Act, paragraphs (c), (d), and (f) of Section 102. Section 102(c) does not allow an applicant to obtain a patent when he "has abandoned the invention." This statute does not refer to disposal of the invention itself, however, but instead to the intentional surrender of an invention *to the public*. Older Supreme Court opinions instruct that abandonment may occur where an inventor expressly dedicates it to the public, through a deliberate relinquishment or conduct evidencing an intent not to pursue patent protection.[91] The circumstances must be such that others could reasonably rely upon the inventor's renunciation.[92] Perhaps because few individuals expressly cede their patentable inventions to the public without seeking compensation, there are few modern judicial opinions that consider 35 U.S.C. § 102(c) in any meaningful way. In addition, the generally applicable principle of equitable estoppel may apparently be used to obtain the same result.[93]

Like section 102(c), section 102(d) of the Patent Act is reportedly little-used.[94] 35 U.S.C. 102(d) bars a U.S. patent when (1) an inventor files a foreign patent application more than twelve months before filing the U.S. application, and (2) a foreign patent results from that application prior to the U.S. filing date. Suppose that an inventor files an application at a foreign patent office on May 25, 2004. The foreign application matures into a granted foreign patent on August 1, 2005. If the inventor has not filed his patent application at the USPTO as of August 1, 2005, the date of the foreign patent grant, any patent application that the inventor subsequently filed in the United States would be defeated.

The policy basis for 35 U.S.C. § 102(d) is to encourage the prompt filing of patent applications in the United States. As the Patent Office Commissioner explained in 1870:

> The intention of [C]ongress obviously was to obtain for this country the free use of the inventions of foreigners as soon as they became free abroad. This is indicated by the use of the phrase, 'first patented, or caused to be patented, in a foreign country,' for it was presumable that American citizens would obtain their first patent here, while a foreigner would first patent his invention in his own country. The statute was designed to prevent a foreigner from spending his time and capital in the development of an invention in his own country, and then coming to this country to enjoy a further monopoly, when the invention had become free at home. The result of such a course would be that while the foreign country was developing the

invention and enjoying its benefits, its use could be interdicted here; while, if the term of the monopoly could be further extended here, the market could be controlled long after the foreign nation was prepared to flood this country with the unpatented products of the patented process.[95]

Section 102(d) has been subject to critical commentary. Because inventors may choose to file a patent application only in the United States, the policy goal of assuring that the U.S. market will become patent-free contemporaneously with foreign markets may not be well-served by this provision. In addition, 35 U.S.C. § 102(d) effectively acts against foreign, rather than U.S.-based inventors, as domestic inventors ordinarily file at the USPTO first before seeking rights overseas. Some commentators have suggested that 35 U.S.C. § 102(d) violates the spirit, if not the letter, of U.S. international treaty obligations, which generally impose an obligation of national treatment with respect to intellectual property matters.[96]

Finally, H.R. 2795 would also eliminate current 35 U.S.C. § 102(f), which states that a person may obtain a patent unless "he did not himself invent the subject matter sought to be patented." As this requirement that only an actual inventor may obtain a patent is also stated by 35 U.S.C. § 101,[97] this proposed amendment does not appear to have any substantive impact upon the patent law.

Assignee Filing

Under current U.S. law, a patent application must be filed by the inventor —that is to say, the natural person or persons who developed the invention.[98] This rule applies even where the invention was developed by individuals in their capacity as employees. Even though rights to the invention have usually been contractually assigned to an employer, for example, the actual inventor, rather than the employer, must be the one that applies for the patent. Section 118 of the Patent Act allows a few exceptions to this general rule. If an inventor cannot be located, or refuses to perform his contractual obligation to assign an invention to his employer, then the employer may file in place of the inventor.[99]

H.R. 2795 would instead stipulate that a "person to whom the inventor has assigned or is under an obligation to assign the invention may make an application for patent."[100] Individuals who otherwise make a showing of a "sufficient proprietary interest in the matter" may also apply for a patent on behalf of the inventor upon a sufficient show of proof of the pertinent facts. Under the proposed legislation, if the USPTO "Director grants a patent on an application filed under this section by a person other than the inventor, the patent shall be granted to the real party in interest and upon such notice to the inventor as the Director considers to be sufficient."

Legal reforms allowing assignee filing of patent applications have been discussed for many years. A 1966 Report of the President's Commission on the Patent System recommended this change as a way to simplify formalities of application filing and of avoiding the delays caused by the need to identify and obtain signatures from each inventor.[101] The 1992 Advisory Commission on Patent Law Reform was also in favor of this change. The 1992 Commission observed that the United States was "the only country which does not permit the assignee of an invention to file a patent application in its own name."[102] In the opinion of the 1992 Commission, assignee filing would appropriately accompany a U.S. shift to a first-inventor-to-file priority system, as the reduction of formalities would allow innovative enterprises to more promptly file patent applications.

The 1992 Commission also reviewed potential undesirable aspects of assignee filing. The Commission noted that patent applications filed by assignees may lack the actual inventor's personal guarantee that the application was properly filed. In addition, assignee filing might derogate the right of natural persons to their inventions. In the opinion of the Commission, however, the advantages of assignee filing outweighed the disadvantages.[103]

Elimination of the Best Mode Requirement

H.R. 2795 would eliminate U.S. patent law's best mode requirement. Currently, inventors are required to "set forth the best mode contemplated by the inventor of carrying out his invention."[104] Failure to disclose the best mode known to the inventor is a ground for invalidating an issued patent. The courts have established a two-part standard for analyzing whether an inventor disclosed her best mode in a particular patent. The first inquiry was whether the inventor knew of a way of practicing the claimed invention that he considered superior to any other. If so, then the patent instrument must identify, and disclose sufficient information to enable persons of skill in the art to practice that best mode.[105]

Proponents of the best mode requirement have asserted that it allows the public to receive the most advantageous implementation of the technology known to the inventor. This disclosure becomes part of the patent literature and may be freely reviewed by those who wish to design around the patented invention. Members of the public are also said to be better able to compete with the patentee on equal footing after the patent expires.[106]

The best mode requirement has encountered severe criticism in recent years, however.[107] For example, a 1992 Presidential Commission recommended that Congress eliminate the best mode requirement. The Commission reasoned that patents are also statutorily required to disclose "the manner and process of making and using [the invention], in such full, clear, concise, and exact terms as to enable any person skilled in the art . . . to make and use the same."[108] This "enablement" requirement was believed to provide sufficient information to achieve the patent law's policy goals.[109]

The Commission further stated that the best mode requirement leads to increases in the costs and complexity of patent litigation. As the Commission explained:

> The disturbing rise in the number of best mode challenges over the past 20 years may serve as an indicator that the best mode defense is being used primarily as a procedural tactic. A party currently can assert failure to satisfy the best mode requirement without any significant burden. This assertion also entitles the party to seek discovery on the "subjective beliefs" of the inventors prior to the filing date of the patent application. This broad authority provides ample opportunity for discovery abuse. Given the fluidity by which the requirement is evaluated (e.g., even accidental failure to disclose any superior element, setting, or step can negate the validity of the patent), and the wide ranging opportunities for discovery, it is almost certain that a best mode challenge will survive at least initial judicial scrutiny.[110]

The Commission further reasoned that the best mode at the time of filing is unlikely to remain the best mode when the patent expires many years later.[111] Because many foreign patent laws include no analog to the best mode requirement, inventors based overseas have also questioned the desirability of the best mode requirement in U.S. law.

Inequitable Conduct

The administrative process of obtaining a patent from the USPTO has traditionally been conducted as an *ex parte* procedure. Stated differently, patent prosecution involves only the applicant and the USPTO. Members of the public, and in particular the patent applicant's marketplace competitors, do not participate in patent acquisition procedures.[112] As a result, the patent system relies to a great extent upon applicant observance of a duty of candor and truthfulness towards the USPTO.

An applicant's obligation to proceed in good faith may be undermined, however, by the great incentive applicants possess not to disclose, or to misrepresent, information that might deleteriously impact their prospective patent rights. The patent law therefore penalizes those who stray from honest and forthright dealings with the USPTO. Under the doctrine of "inequitable conduct," if an applicant intentionally misrepresents a material fact or fails to disclose material information, then the resulting patent will be declared unenforceable.[113] Two elements must exist before a court will decide that the applicant has engaged in inequitable conduct. First, the patentee must have misrepresented or failed to disclose material information to the USPTO in the prosecution of the patent.[114] Second, such nondisclosure or misrepresentation must have been intentional.[115]

During patent infringement litigation, an accused infringer has the option of asserting that the plaintiff's patent is unenforceable because it was procured through inequitable conduct. Concerns have arisen that charges of inequitable conduct have become routine in patent cases. As one commentator explains:

> The strategic and technical advantages that the inequitable conduct defense offers the accused infringer make it almost too attractive to ignore. In addition to the potential effect on the outcome of the litigation, injecting the inequitable conduct issue into patent litigation wreaks havoc in the patentee's camp. The inequitable conduct defense places the patentee on the defensive, subjects the motives and conduct of the patentee's personnel to intense scrutiny, and provides an avenue for discovery of attorney-client and work product documents . . . . [116]

As the Federal Circuit put it, "the habit of charging inequitable conduct in almost every major patent case has become an absolute plague."[117] Other observers believe that because inequitable conduct requires an analysis of the knowledge and intentions of the patent applicants, the doctrine may also be contributing disproportionately to the time and expense of patent litigation.[118]

Due to these perceived burdens upon patent litigation, some commentators have proposed that the inequitable conduct defense be eliminated.[119] Others believe that inequitable conduct is necessary to ensure the proper functioning of the patent system. As the Advisory Commission on Patent Law Reform explained in its 1992 report:

Some mechanism to ensure fair dealing between the patentee, public, and the Federal Government has been part of the patent system for over 200 years. In its modern form, the unenforceability defense provides a necessary incentive for patent applicants to engage in fair and open dealing with the [USPTO] during the ex parte prosecution of patent applications, by imposing the penalty of forfeiture of patent rights for failure to so deal. The defense is also considered to be an essential safeguard against truly fraudulent conduct before the [USPTO]. Finally, the defense provides a means for encouraging complete disclosure of information relevant to a particular patent application. . . . . Thus, from a policy perspective, the defense of unenforceability based upon inequitable conduct is desirable and should be retained.[120]

Proposed legislation would retain the concept of an inequitable conduct defense, but introduce a number of substantive and procedural changes to the doctrine. In broad overview, H.R. 2795 would provide statutory authorization for the USPTO Director to issue regulations governing applicants' duty of candor.[121] Although the USPTO has for many years issued such regulations,[122] commentators have questioned the authority of the USPTO to do so under current law.[123]

H.R. 2795 would also provide the USPTO with authority to prosecute violations of the inequitable conduct doctrine. Although that agency previously performed this function without express statutory authorization, the USPTO suspended that activity in 1992.[124] Under H.R. 2795, the USPTO would be obligated to establish a "special office with authority to investigate possible violations of the duty of candor and good faith." The USPTO would be granted subpoena power and be able to conduct hearings in pursuit of these investigations. The USPTO would also be able to penalize misconduct through substantial civil fines.

In addition, H.R. 2795 would limit the circumstances under which the defense of inequitable conduct could be raised before the courts. In broad outline, under H.R. 2795 if a court determines an issue of possible misconduct, then the court is directed to refer the matter to the USPTO. Within judicial infringement proceedings, issues of inequitable conduct could only arise after the court grants a motion to amend the pleadings. Such a motion would have to describe the relevant facts in detail and could not be granted until the court has previously entered a judgment that at least one of the asserted patent claims is invalid.

H.R. 2795 would also introduce a substantive change to the law of inequitable conduct. Under this legislation, a finding of inequitable conduct is appropriate only in circumstances when the USPTO examiner relied upon the alleged misconduct. In particular, the charge of inequitable conduct cannot be sustained unless the USPTO "would not have issued the invalidated claim, acting reasonably, in the absence of the misconduct," or "based upon the prosecution history as a whole objectively considered, would have done so based upon in whole or in part on account of the misconduct."[125]

Publication of Pending Applications

Until recent years, the U.S. patent system maintained pending patent applications in secrecy. The first moment that the public would become aware of the existence of a U.S. patent application was the day the USPTO formally allowed it to issue as a granted patent. This regime advantaged patent applicants because it allowed them to understand exactly what the scope of any allowed claims might be prior to disclosing an invention. Thus, if the applicant was able to maintain the invention that was subject to a patent application as a trade secret, then he could choose between obtaining the allowed patent claims and trade secret status. In addition, because the invention was not disclosed prior to the award of formal patent rights, unscrupulous competitors were discouraged from copying the invention.

However, this secrecy regime has been perceived as imposing costs as well. Others might well engage in duplicative research efforts during the pendency of patent applications, unaware that an earlier inventor had already staked a claim to that technology. This arrangement also allowed inventors to commence infringement litigation on the very day a patent issued, without any degree of notice to other members of the technological community.[126]

Industry in the United States possessed one mechanism for identifying pending U.S. patent applications. Most foreign patent regimes publish all pending patent applications approximately 18 months after they have been filed.[127] As a result, savvy firms in the United States could review pending applications filed before foreign patent offices, and make an educated guess as to the existence of a corresponding U.S. application. This effort was necessarily inexact, however, particularly as some inventors either lacked the resources, or made the strategic decision, not to obtain patent rights outside the United States.

In enacting the American Inventors Protection Act of 1999, Congress for the first time introduced the concept of pre-grant publication into U.S. law. Since November 29, 2000, U.S. patent applications have been published 18 months from the date of filing, with some exceptions. The most significant of these exceptions applies where the inventor represents that he will not seek patent protection abroad. In particular, if an applicant certifies that the invention disclosed in the U.S. application will not be the subject of a patent application in another country that requires publication of applications 18 months after filing, then the USPTO will not publish the application.[128] As a result, inventors who do not wish to seek foreign patent rights retain the possibility of avoiding pre-grant publication.

Proposed legislation would further modify the U.S. pre-grant publication system by effectively calling for all pending applications to be published approximately 18 months after they are filed. In particular, H.R. 2795 would eliminate the possibility of opting out of pre-grant publication by certifying that a patent will be sought only in the United States.[129]

Pre-Issuance Submissions

Pending legislation would expand the ability of members of the public to submit information to the USPTO that is pertinent to pending applications. Under current law, interested individuals may enter a protest against a patent application. The protest must specifically identify the application and be served upon the applicant. The protest must also include a copy and, if necessary, an English translation, of any patent, publication or other information relied upon. The protester also must explain the relevance of each item.[130]

Protest proceedings have traditionally played a small role in U.S. patent practice. Until Congress enacted the American Inventors Protection Act of 1999, the USPTO maintained applications in secrecy. Therefore, the circumstances in which members of the public would learn of the precise contents of a pending patent application were relatively limited. With the USPTO commencing publication of some pending patent applications, protests would seem far more likely. Seemingly aware of this possibility, the 1999 Act provided that the USPTO shall "ensure that no protest or other form of pre-issuance opposition . . . may be initiated after publication of the application without the express written consent of the applicant."[131] Of course, the effect of this provision is to eliminate the possibility of protest in exactly that class of cases where the public is most likely to learn of the contents of a pending application.

Through rulemaking, the USPTO has nonetheless established a limited mechanism for members of the public to submit information they believe is pertinent to a pending, published application. The submitted information must consist of either a patent or printed publication, and it must be submitted within two months of the date the USPTO published the pending application. Nondocumentary information that may be relevant to the patentability determination, such as sales or public use of the invention, will not be considered.[132] In addition, because Congress stipulated that no protest or pre-grant opposition may occur absent the consent of the patent holder, the USPTO has explained that it will not accept *comments* or *explanations* concerning the submitted patents or printed publications. If such comments are attached, USPTO staff will redact them before the submitted documents are forwarded to the examiner.[133]

Pending legislation would augment the possibility for pre-issuance submissions. Under H.R. 2795, any person would be able to submit patent documents and other printed publications to the USPTO for review. Such prior art must be submitted within the later date of either (1) the date the USPTO issues a notice of allowance to the patent applicant; or (2) either six months after the date of pre-grant publication of the application, or the date of the rejection of any claim by the USPTO examiner. Such a submission must include "a concise description of the asserted relevance of each submitted document."[134]

Most observers agree that ideally, the USPTO would have access to all pertinent information when making patentability determinations. A more expansive pre-issuance submission policy may allow members of the public to disclose relevant patents and other documents that the USPTO's own searchers may not have revealed, thereby leading to more accurate USPTO decision making. On the other hand, lengthy pre-issuance submissions may merely be repetitive of the USPTO's own search results, but still require extensive periods of examiner review that might ultimately delay examination. The proposal of H.R. 2795 apparently attempts to balance these concerns by expanding existing opportunities for post-publication submissions, but limiting the timing and nature of those submissions so as to prevent undue burdens upon the USPTO and patent applicants.

Continuation Applications

Pending legislation would allow the USPTO Director to limit the availability of so-called continuation applications via regulation. Continuation applications essentially allow inventors to re-file previously rejected patent applications in order to pursue further prosecution with a USPTO examiner.[135] The filing of a patent application effectively allows two formal communications, termed "Office Actions," with a USPTO examiner. Agreement often cannot be reached by this point, however, leaving the applicant with only the alternatives of abandonment of patent protection or the filing of an appeal. By filing a continuation application, an applicant essentially purchases additional time for dialogue between the applicant and examiner.

The use of continuation applications is commonplace in U.S. patent practice. Applicants not infrequently file one or more continuing applications based upon an earlier filed "patent" application. Many patents have issued based upon chains of continuation applications involving a parent, grandparent, and even more remote predecessors.

Continuation applications are said to allow applicants to more accurately claim a previously disclosed invention without the necessity of an appeal.[136] Some commentators believe they are subject to abuse, however. Under this view, continuation practice introduces delay and uncertainty into the patent acquisition process. In particular, applicants are said to use a chain of continuation applications in order to gain advantages over competitors by waiting to see what product the competitor will make, and then drafting patent claims that cover that product. Continuation practice is also said to have led to long delays in the issuance of a patent in order to surprise an established industry, a process known as "submarine patenting."[137]

At present time, two features of the Patent Act have been identified as mitigating against abuse of continuation applications. The first feature, discussed above, is the pre-grant publication of pending patent applications.[138] This rule allows interested members of the public to be notified that the application has been filed at the USPTO prior to the formal bestowal of exclusive patent rights. Under current law, however, exceptions exist to this rule, and as a result some applications are not published prior to their issuance.[139]

The second feature is the term of the patent, which as of June 8, 1995, was set to the duration of twenty years from the date of filing.[140] The twenty-year term is measured from the filing date of the earliest in the chain of continuation applications. Inventors who file a series of continuation applications thereby shorten their term of eventual patent protection. In some circumstances, however, even a patent with a diminished term may remain of competitive significance. And again there is an exception to the twenty-year patent term based upon the filing date: Applications filed prior to June 8, 1995, may instead enjoy a term of seventeen years set from the date of issuance.

H.R. 2795 would allow the USPTO Director to issue regulations that would limit the ability of inventors to file continuation applications. The bill would further stipulate that "[n]o such regulation may delay applicants an adequate opportunity to obtain claims for any invention disclosed in an application for patent."[141]

Prior User Rights

Pending legislation would expand the applicability of a "first inventor defense" established by the American Inventors Protection Act of 1999. As currently found at 35 U.S.C. § 273, an earlier inventor of a "method of doing or conducting business" that was later patented by another may claim a defense to patent infringement in certain circumstances. H.R. 2795 would broaden this defense by allowing it to apply with respect to any patented subject matter.

The impetus for this provision lies in the rather complex relationship between the law of trade secrets and the patent system. Trade secrecy protects individuals from misappropriation of valuable information that is useful in commerce. One reason an inventor might maintain the invention as a trade secret rather than seek patent protection is that the subject matter of the invention may not be regarded as patentable. Such inventions as customer lists or data compilations have traditionally been regarded as amenable to trade secret protection but not to patenting.[142] Inventors might also maintain trade secret protection due to ignorance of the patent system or because they believe they can keep their invention as a secret longer than the period of exclusivity granted through the patent system.[143]

The patent law does not favor trade secret holders, however. Well-established patent law provides that an inventor who makes a secret, commercial use of an invention for more than one year prior to filing a patent application at the USPTO forfeits his own right to a patent.[144] This policy is based principally upon the desire to maintain the integrity of the statutorily prescribed patent term. The patent law grants patents a term of twenty years, commencing from the date a patent application is filed.[145] If the trade secret holder could make commercial use of an invention for many years before choosing to file a patent application, he could disrupt this regime by delaying the expiration date of his patent.

On the other hand, settled patent law principles established that prior secret uses would not defeat the patents of later inventors.[146] If an earlier inventor made secret commercial use of an invention, and another person independently invented the same technology later and obtained patent protection, then the trade secret holder could face liability for patent infringement. This policy is based upon the reasoning that once issued, published patent instruments fully inform the public about the invention, while trade secrets do not. As between a subsequent inventor who patented the invention, and thus had disclosed the invention to the public, and an earlier trade secret holder who had not, the law favored the patent holder.

An example may clarify this rather complex legal situation. Suppose that Inventor A develops and makes commercial use of a new manufacturing process. Inventor A chooses not to obtain patent protection, but rather maintains that process as a trade secret. Many years later, Inventor B independently develops the same manufacturing process and promptly files a patent application claiming that invention. In such circumstances, Inventor A's earlier, trade secret use does not prevent Inventor B from procuring a patent. Furthermore, if the USPTO approves the patent application, then Inventor A faces infringement liability should Inventor B file suit against him.

The American Inventors Protection Act of 1999 somewhat modified this principle. That statute in part provided an infringement defense for an earlier inventor of a "method of doing or conducting business" that was later patented by another. By limiting this defense to patented methods of doing business, Congress responded to the 1998 Federal Circuit opinion in *State Street Bank and Trust Co. v. Signature Financial Group*.[147] That judicial opinion recognized that business methods could be subject to patenting, potentially exposing individuals who had maintained business methods as trade secrets to liability for patent infringement.

Again, an example may aid understanding of the first inventor defense. Suppose that Inventor X develops and exploits commercially a new method of doing business. Inventor X maintains his business method as a trade secret. Many years later, Inventor Y independently develops the same business method and promptly files a patent application claiming that invention. Even following the enactment of the American Inventors Protection Act, Inventor X's earlier, trade secret use would not prevent Inventor Y from procuring a patent. However, should the USPTO approve Inventor Y's patent application, and should Inventor Y sue Inventor X for patent infringement, then Inventor X may potentially claim the benefit of the first inventor defense. If successful,[148] Inventor X would enjoy a complete defense to infringement of Inventor Y's patent.

As originally enacted, the first inventor defense applied only to patents claiming a "method of doing or conducting business." Although the American Inventors Protection Act did not define this term, the first inventor defense was arguably a focused provision directed towards a specific group of potential patent infringers.

Pending legislation would expand upon the first inventor defense by allowing it to apply to all patented subject matter.[149] By removing current restrictions referring to methods of doing business, H.R. 2795 would effectively introduce "prior user rights" into U.S. law.

A feature of many foreign patent regimes, prior user rights are often seen as assisting small entities, which may lack the sophistication or resources to pursue patent protection. The provision of prior user rights would allow such entities to commercialize their inventions when they used the subject matter of the invention prior to the patent's filing date, even when they themselves did not pursue patent rights. For this reason, a more expansive prior user rights regime has also been tied to adoption of the first-inventor-to-file priority system.[150]

Proponents of prior user rights also assert that the new legislation would support investment in technological innovation. Under this view, firms would not longer be required to engage in extensive defensive patenting, but rather would be able to devote these resources to further innovation. In addition, some commentators observe that many U.S. trading partners, including Germany and Japan, currently allow prior user rights. As a result, U.S. firms that obtain patent rights in certain foreign nations may face the possibility that a foreign firm may enjoy prior user rights in that invention. Foreign firms with U.S. patents do not currently face this possibility with respect to U.S. firms, however. Under this view, adoption of prior user rights in the United States would "level the playing field" for U.S. industry.[151]

Proposals to adopt prior user rights have attracted critics, however. Some observers believe that this regime would benefit large corporations at the expense of smaller ones. Others believe that individuals who are aware that they can rely upon prior user rights will be less likely to disclose their inventions through the patent system. Still others have stated that prior user rights reduce the value of patents and therefore make innovation less desirable. The role of the U.S. Constitution is sometimes debated within this context as well. Article I, section 8, clause 8 of the Constitution provides Congress with the authority to award "inventors the exclusive right to their . . . discoveries." Some commentators suggest this language suggests, or possibly requires, a system of exclusive patent rights, rather than an interest that may be mitigated by prior user rights.[152]

Injunctions

35 U.S.C. § 283 currently allows courts to "grant injunctions in accordance with the principles of equity to prevent the violation of any right secured by patent, on such terms as the court deems reasonable."[153] H.R. 2795 would add the following additional language to that statute:

> In determining equity, the court shall consider the fairness of the remedy in light of all the facts and the relevant interests of the parties associated with the invention. Unless the injunction is entered pursuant to a nonappealable judgment of infringement, a court shall stay

the injunction pending an appeal upon an affirmative showing that the stay would not result in irreparable harm to the owner of the patent and that the balance of the hardships from the stay does not favor the owner of the patent.[154]

In other words, this legislation would stipulate that all the facts and relevant interests must be considered when a court weighs "principles of equity" in deciding whether to enjoin an adjudicated infringer or not.

Understanding of this proposed reform requires a review of current judicial precedent concerning the availability of injunctions in patent cases. Under current law, courts ordinarily grant permanent injunctions to patentees that prevail in infringement litigation.[155] As the Federal Circuit recently explained:

> Because the "right to exclude recognized in a patent is but the essence of the concept of property," the general rule is that a permanent injunction will issue once infringement and validity have been adjudged. To be sure, "courts have in rare instances exercised their discretion to deny injunctive relief in order to protect the public interest." *Rite-Hite Corp. v. Kelley, Inc.*, 56 F.3d 1538, 1547 (Fed. Cir.1995); see *Roche Prods., Inc. v. Bolar Pharm. Co.*, 733 F.2d 858, 865-66 (Fed. Cir.1984) ("standards of the public interest, not the requirements of private litigation, measure the propriety and need for injunctive relief"). Thus, we have stated that a court may decline to enter an injunction when "a patentee's failure to practice the patented invention frustrates an important public need for the invention," such as the need to use an invention to protect public health.[156]

As this language suggests, few published judicial opinions decline to grant injunctions against adjudicated patent infringers. The usually cited exception is the 1934 decision in *City of Milwaukee v. Activated Sludge*,[157] where the Court of Appeals for the Seventh Circuit refused to enjoin infringement of a patented method for sewage treatment. Had the city of Milwaukee been prevented from using the patented invention, it would have been required to dump large quantities of raw sewage into Lake Michigan. Observing that "the health and the lives of more than half a million people are involved,"[158] the court denied the requested injunction. Because the patentee still obtained judicially determined monetary remedies against the city of Milwaukee, this outcome essentially amounted to an award of a compulsory license.

Another notable case, the 1944 decision in *Vitamin Technologists, Inc. v. Wisconsin Alumni Research Foundation*,[159] involved a patent claiming a method of irradiating foods to increase Vitamin D content. This treatment helped eliminate the debilitating Vitamin D deficiency disease called rickets. The availability of Vitamin D-enhanced margarine was particularly important to the poor, who were better able to afford margarine as compared to butter. The patent proprietor had nonetheless refused to license the patent to margarine producers. The Court of Appeals for the Ninth Circuit ultimately held the asserted patents invalid or unenforceable on conventional grounds. However, the court also discussed the concept that injunctions should be refused where they act against public health interests.[160]

Other courts have taken a different route by delaying the effective date of a permanent injunction rather than refusing it altogether. For example, in *Schneider (Europe) AG v. SciMed Life Systems, Inc.*,[161] the adjudicated infringer marketed a rapid-exchange catheter used by surgeons. Although the court concluded that no evidence of record supported a finding that the infringing product was more safe or objectively superior to other catheters on the market, the court recognized that some physicians did strongly prefer the infringing product.[162] The court opted to grant a permanent injunction with a delay of one year from the entry of judgment. The court reasoned that this year-long transition period would allow surgeons to switch from the infringing product with a minimum of disruption, at least in comparison with the immediate imposition of an injunction. The court further provided that the patentee would receive an escalating royalty during the transition period.[163]

Some commentators have expressed concerns over the current state of the law. It has been observed that although the patent statute directs courts to "grant injunctions in accordance with the principles of equity,"[164] in practice courts issue injunctions in favor of victorious patent proprietors "virtually automatically."[165] In the view of some observers, this rule has encouraged strategic behavior by speculators. These speculators do not themselves market goods or services, and thus do not acquire patents in order to protect their own markets. Rather, they are said to use patents to threaten manufacturers and service providers. Because such speculators may legitimately threaten to halt use of the patented invention entirely, accused infringers may enter into a license on even a dubious patent in the face of losing the company business. This practice is sometimes termed "trolling," after creatures from folklore that would emerge from under a bridge in order to waylay travelers.[166]

In response, some commentators have observed that not all patentees that chose not to market their patented inventions are fairly characterized as opportunistic speculators. Particular individuals or firms may lack the expertise or resources to produce a patented product, or otherwise have sound business reasons not to do so. In addition, as the Federal Circuit explained a decade ago:

> A patent is granted in exchange for a patentee's disclosure of an invention, not for the patentee's use of the invention. There is no requirement in this country that a patentee make, use, or sell its patented invention.[167]

Still other observers view patents as time-limited property rights.[168] Under this view, infringers are properly enjoined so that the patent owner's exclusive rights may be preserved. Failing to enjoin infringements may also diminish the incentives needed for innovators to develop patentable inventions in the first instance.

On the other hand, scholarly commentary concludes that the term "intellectual property" is one of relatively recent vintage, and that patents can just as readily be analogized as arising under the tort law, or as constituting a government-granted subsidy.[169] Although viewing a patent as a property right arguably suggests that an injunction is a suitable remedy for a violation of that right, these alternative conceptions of the patent grant do not so strongly imply that courts should enjoin adjudicated infringers as a matter of course. It should also be noted that even traditional properties, such as real estate, are prospectively subject to numerous limitations, including easements, zoning restrictions, servitudes, and other obligations.[170]

Limitation Upon Damages

Pending legislation would also address the award of damages where the patented invention forms but one component of the infringer's larger commercial product or process. In the wording of H.R. 2795:

> In determining a reasonable royalty in the case of a combination, the court shall consider, if relevant and among other factors, the portion of the realizable profit that should be credited to the inventive contribution as distinguished from other features of the combination, the manufacturing process, business risks, or significant features or improvements added by the infringer.[171]

This proposed reform appears to be directed towards perceived concerns about overly generous damages awards in this context. As one commentator asserted:

> [I]nventors have learned to abuse the patent system and increase leverage against manufacturers by pursuing "system claims" in the [USPTO]. These clever claims insert the crux of the predator's "innovation" into much larger contexts than that to which the inventor is rightfully entitled. For example, the abuser may actually have invented a hinge mechanism, but draws the patent claim to a door including the hinge mechanism. In this example, the door is well known to, and long in use by, the public but in subsequent litigation, the patent predator claims entitlement to, and the court awards, damages based on the entire value of the door rather than on the value of the innovative hinge.[172]

Several observations may be made about this proposed reform. First, the Federal Circuit has explained that:

> Virtually *all* patents are "combination patents," if by that label one intends to describe patents having claims to inventions formed of a combination of elements. It is difficult to visualize, at least in the mechanical-structural arts, a "non-combination" invention, i.e., an invention consisting of a *single* element. Such inventions, if they exist, are rare indeed.[173]

Under this view, this legislative proposal is of potentially broad application. Further, the required assessment of the "inventive contribution" of a patented combination —rather than base its damages determination upon the claims themselves — would mark a notable change in U.S. patent law. As the Federal Circuit recently stated:

It is well settled that "there is no legally recognizable or protected 'essential' element, gist or 'heart' of the invention in a combination patent." Rather, "'[t]he invention' is defined by the claims."[174]

Finally the language of H.R. 2795 expressly applies only to an award of a reasonable royalty. However, the patent statute also authorizes the courts to award damages equal to the lost profits of the patentee,[175] provided the patentee can make the necessary showing. It is unclear why this proposal is limited to damages awards based upon reasonable royalties, rather than upon both forms of damages awards available under the patent laws.

Willful Infringement

Pending legislation would also reform the law of willful infringement. The patent statute currently provides that the court "may increase the damages up to three times the amount found or assessed."[176] An award of enhanced damages, as well as the amount by which the damages will be increased, is committed to the discretion of the trial court.[177] Although the statute does not specify the circumstances in which enhanced damages are appropriate, the courts most commonly award them when the infringer acted in blatant disregard of the patentee's rights. This circumstance is termed "willful infringement."[178]

Courts will not ordinarily enhance damages due to willful infringement if the adjudicated infringer did not know of the patent until charged with infringement in court, or if the infringer acted with the reasonable belief that the patent was not infringed or that it was invalid. Federal Circuit decisions emphasize the duty of someone with actual notice of a competitor's patent to exercise due care in determining if his acts will infringe that patent. A common way to fulfill this obligation is to obtain competent legal advice before commencing, or continuing, activity that may infringe another's patent.[179]

Prior to 2004, the Federal Circuit held that when an accused infringer invoked the attorney-client or work-product privilege, courts should be free to reach an adverse inference that either (1) no opinion had been obtained or (2) an opinion had been obtained and was contrary to the infringers's desire to continue practicing the patented invention.[180] However, in its decision in *Knorr-Bremse Systeme fuer Nutzfahrzeuge GmbH v. Dana Corp.*,[181] the Federal Circuit expressly overturned this principle. The Court of Appeals further stressed that the failure to obtain legal advice did not occasion an adverse inference with respect to willful infringement either. Following the *Knorr-Bremse* opinion, willful infringement determinations are based upon "the totality of circumstances, but without the evidentiary contribution or presumptive weight of an adverse inference that any opinion of counsel was or would have been unfavorable."[182]

Patent law's willful infringement doctrine has proven controversial. Some observers believe that this doctrine ensures that patent rights will be respected in the marketplace. Critics of the policy believe that the possibility of trebled damages discourages individuals from reviewing issued patents. Out of fear that their inquisitiveness will result in multiple damages, innovators may simply avoid looking at patents until they are sued for infringement. To the extent this observation is correct, the law of willful infringement discourages the dissemination of technical knowledge, thereby thwarting one of the principal goals of the patent system. Fear of increased liability for willful infringement may also discourage firms from challenging patents of dubious validity. Consequently some have argued that the patent system should shift to a "no-fault" regime of strictly compensatory damages, without regard to the state of mind of the adjudicated infringer.[183]

H.R. 2795 would add several clarifications and changes to the law of willful infringement. First, a finding of willful infringement would be appropriate only where (1) the infringer received specific written notice from the patentee and continued to infringe after a reasonable opportunity to investigate; (2) the infringer intentionally copied from the patentee with knowledge of the patent; and (3) the infringer continued to infringe after an adverse court ruling. Second, willful infringement cannot be found where the infringer possessed an informed, good faith belief that its conduct was not infringing. Finally, a court may not determine willful infringement before the date on which a determination is made that the patent is not invalid, enforceable, and infringed.[184]

Post-Issuance Opposition Proceedings

Pending legislation would introduce post-issuance opposition proceedings into U.S. patent law. Oppositions, which are common in foreign patent regimes, are patent revocation proceedings that are usually administered by authorities from the national patent office. Oppositions often involve a wide range of potential invalidity arguments and are conducted through adversarial hearings that resemble courtroom litigation.

Although the U.S. patent system does not currently include oppositions, the U.S. patent system has incorporated a so-called reexamination proceeding since 1981. Some commentators have viewed the reexamination as a more limited form of an opposition. Under the reexamination statute, any individual, including the patentee, a competitor, and even the USPTO Director, may cite a prior art patent or printed publication to the USPTO. If the USPTO determines that this reference raises a "substantial new question of patentability" with respect to an issued patent, then it will essentially reopen prosecution of the issued patent.

Traditional reexamination proceedings are conducted in an accelerated fashion on an *ex parte* basis. Following the American Inventors Protection Act of 1999, an *inter partes* reexamination allows the requester to participate more fully in the proceedings through the submission of arguments and the filing of appeals. Either sort of reexamination may result in a certificate confirming the patentability of the original claims, an amended patent with narrower claims or a declaration of patent invalidity.

Congress intended reexamination proceedings to serve as an inexpensive alternative to judicial determinations of patent validity.[185] Reexamination also allows further access to the legal and technical expertise of the USPTO after a patent has issued.[186] However, some commentators believe that reexamination proceedings have been employed only sparingly and question their effectiveness.[187]

Some analysts have expressed concern that potential requesters are discouraged from commencing *inter partes* reexamination proceedings due to a statutory provision that limits their future options. In order to discourage abuse of these proceedings, the *inter partes* reexamination statute provides that third-party participants may not later assert that a patent is invalid "on any ground that [they] raised or could have raised during the inter partes reexamination proceedings."[188] Some observers believe that this potential estoppel effect disinclines potential requesters from use of this post-issuance proceeding. In apparent response to this concern, H.R. 2795 would delete the phrase "or could have raised" from the statute.[189] As a result, *inter partes* reexamination requesters would be limited only with respect to arguments that they actually made before the USPTO.

In addition to reforming the reexamination statute, H.R. 2795 would create an additional post-issuance proceeding termed an "opposition." Under this bill, any person could commence an opposition either within nine months after the issuance of the patent, or six months after receiving notice from the patent holder alleging infringement. The opposition may relate to a wide range of patentability issues, including double patenting, statutory subject matter, novelty, nonobviousness, enablement, and definite claiming. The commencement of the opposition is conditioned upon the USPTO Director's determination that the opponent has raised a substantial question of patentability with respect to at least one claim in the patent.

H.R. 2795 further provides that opposition proceedings will be tried before a panel of three administrative patent judges. In the event that the patentee files an infringement suit within nine months of patent issuance, or six months of notifying the alleged infringer, the opposition will be stayed upon the request of the patent owner. The patent owner may amend its claims during an opposition, provided that those amendments do not broaden the scope of the claims. The opposition must conclude within one year of its commencement, although one six-month extension is possible. The results of opposition proceedings may be appealed to the courts.

Many observers have called for the United States to adopt an opposition system in order to provide more timely, lower cost, and more efficient review of issued patents.[190] Such a system could potentially improve the quality of issued patents by weeding out invalid claims. It might also encourage innovative firms to review issued patents soon after they are granted, thereby increasing the opportunity for technology spillovers.[191] Concerns have arisen over oppositions because they too may be costly, complex, and prone to abuse as a means for harassing patent owners.[192] A successful opposition proceeding would require a balancing of these concerns.

CONCLUDING OBSERVATIONS

Pending legislation would arguably work the most sweeping reforms to the U.S. patent system since the nineteenth century. However, many of these proposals, such as pre-issuance publication, prior user rights, and oppositions, have already been implemented in U.S. law to a more limited extent. These and other proposed modifications, such as the first-inventor-to-file priority system and elimination of the best mode requirement, also reflect the decades-old patent practices of Europe, Japan, and our other leading trading partners. As well, many of these changes enjoy the support of diverse institutions, including the Federal Trade Commission, National Academies, economists, industry representatives, attorneys, and legal academics.

Other knowledgeable observers are nonetheless concerned that certain of these proposals would weaken the patent right, thereby diminishing needed incentives for innovation. Some also believe that changes of this magnitude, occurring at the same time, do not present the most prudent course for the patent system. Patent reform therefore confronts Congress with difficult legal, practical, and policy issues, but also with the apparent possibility for altering and potentially improving the legal regime that has long been recognized as an engine of innovation within the U.S. economy.

REFERENCES

[1] Statistics from the United States Patent and Trademark Office (USPTO) support this account. In 1980, the USPTO received 104,329 utility patent applications; by 2003, this

[1]   number had grown to 342,441 applications. During the same time period, the number of U.S. patents granted on an annual basis grew from 61,810 to 169,028. U.S. Patent and Trademark Office, *U.S. Patent Statistics, Calendar Years 1863 - 2003* [available at [http://www.uspto.gov]].
[2] National Research Council, National Academy of Sciences, *A Patent System for the 21st Century*, [Washington, National Academies Press, 2004] and Federal Trade Commission, To Promote Innovation: The Proper Balance of Competition and Patent Law and Policy, October 2003, available at [http://www.ftc.gov].
[3] P.L. 82-593, 66 Stat. 792 (codified at Title 35 United States Code).
[4] 35 U.S.C. § 111.
[5] 35 U.S.C. § 131.
[6] John R. Thomas, "On Preparatory Texts and Proprietary Technologies: The Place of Prosecution Histories in Patent Claim Interpretation," 47 *UCLA Law Review* (1999), 183.
[7] 35 U.S.C. § 112.
[8] Ibid.
[9] 35 U.S.C. § 101.
[10] 35 U.S.C. § 102.
[11] 35 U.S.C. § 103.
[12] 35 U.S.C. § 271(a).
[13] 35 U.S.C. § 283.
[14] 35 U.S.C. § 284.
[15] 35 U.S.C. § 154(a)(2). Although patent term is based upon the filing date, the patentee gains no enforceable legal rights until the USPTO allows the application to issue as a granted patent. A number of Patent Act provisions may modify the basic 20-year term, including examination delays at the USPTO and delays in obtaining marketing approval for the patented invention from other federal agencies.
[16] 35 U.S.C. § 282.
[17] 28 U.S.C. § 1295(a)(1).
[18] 28 U.S.C. § 1254(1).
[19] *See* Rebecca S. Eisenberg, "Patents and the Progress of Science: Exclusive Rights and Experimental Use," 56 *University of Chicago Law Review* 1017 (1989).
[20] 35 U.S.C. § 112.
[21] 35 U.S.C. § 154.
[22] Eisenberg, *supra,* at 1017.
[23] Robert P. Merges, "Intellectual Property and the Costs of Commercial Exchange: A Review Essay," 93 *Michigan Law Review* (1995), 1570.
[24] David D. Friedman *et al.*, "Some Economics of Trade Secret Law," 5 *Journal of Economic Perspectives* (1991), 61.
[25] See Frederic M. Sherer, Industrial Market Structure and Economic Performance (1970), 384-87.
[26] *See* John R. Thomas, "Collusion and Collective Action in the Patent System: A Proposal for Patent Bounties," *University of Illinois Law Review* (2001), 305.
[27] Ibid.
[28] *See* Dan L. Burk and Mark A. Lemley, "Is Patent Law Technology-Specific?," 17 *Berkeley Technology Law Journal* (2002), 1155.

[29] CRS Report RL31281, Patent Quality and Public Policy: Issues for Innovative Firms in Domestic Markets, by John R. Thomas.

[30] See Joseph Farrell and Robert P. Merges, "Incentives to Challenge and Defend Patents: Why Litigation Won't Reliably Fix Patent Office Errors and Why Administrative Patent Review Might Help," 19 *Berkeley Technology Law Journal* (2004), 943.

[31] See Bronwyn H. Hall and Dietmar Harhoff, "Post-Grant Reviews in the U.S. Patent System — Design Choices and Expected Impact," 19 *Berkeley Technology Law Journal* (2004), 989.

[32] See Robert P. Merges, "As Many As Six Impossible Patents Before Breakfast: Property Rights for Business Concepts and Patent System Reform," 14 *Berkeley Technology Law Journal* (1999), 577.

[33] See Craig Allen Nard, "Certainty, Fence Building and the Useful Arts," 74 *Indiana Law Journal* (1999), 759.

[34] Mark A. Lemley, "Rational Ignorance at the Patent Office," 95 *Northwestern University Law Review* (2001), 1495.

[35] Steven J. Elleman, "Problems in Patent Litigation: Mandatory Mediation May Provide Settlement and Solutions," 12 *Ohio State Journal on Dispute Resolution* (1997), 759.

[36] Dee Gill, "Defending Your Rights: Protecting Intellectual Property is Expensive," *Wall Street Journal* (25 Sep. 2000), 6.

[37] Mark H. Webbink, "A New Paradigm for Intellectual Property Rights in Software," 2005 Duke Law and Technology Review (May 1, 2005), 15.

[38] See Sarah Lai Stirland, "Will Congress Stop High-Tech Trolls?," *National Journal* (Feb. 26, 2005), 612.

[39] "Microsoft Advocates for Patent Reform," *eWEEK* (March 10, 2005).

[40] Stirland, *supra*, at 613.

[41] CRS Report RL31132, Multinational Patent Acquisition and Enforcement: Public Policy Challenges and Opportunities for Innovative Firms, by John R. Thomas.

[42] See Elizabeth D. Ferrill, "Patent Investment Trusts: Let's Build a Pit to Catch the Patent Trolls," 6 *North Carolina Journal of Law and Technology* (2005), 367.

[43] See Lorraine Woellert, "A Patent War Is Breaking Out on the Hill," *BusinessWeek* 45 (July 4, 2005).

[44] Nicholas Varchaver, "The Patent King," *Fortune* (May 14, 2001), 202.

[45] Michael Ravnitsky, "More Lemelson Suits," *The National Law Journal* (Dec. 17, 2001), B9.

[46] James Pooley, "Opinion: U.S. patent reform-a good invention," *Electronic Business* (1 Jan. 2000), 72.

[47] See Ronald J. Mann, "Do Patents Facilitate Financing in the Software Industry?," 83 Texas Law Review (2005), 961.

[48] See David G. Barker, "Troll or No Troll? Policing Patent Usage with An Open Post-Grant Review," 2005 *Duke Law and Technology Review* (Apr. 15, 2005), 11.

[49] 35 U.S.C. § 112.

[50] National Science Board, *Science and Engineering Indicators, 1993* (Dec. 8, 1993), 185. See also CRS Report RL30216, Small, High Tech Companies and Their Role in the Economy: Issues in the Reauthorization of the Small Business Innovation (SBIR) Program, by Wendy H. Schacht.

[51] For example, the National Academy of Engineering concluded that "small high-tech companies play a critical and diverse role in creating new products and services, in developing new industries, and in driving technological change and growth in the U.S. economy." National Academy of Engineering, *Risk and Innovation: The Role and Importance of Small High-Tech Companies in the U.S. Economy* (Washington: National Academy Press, 1995), 37. This assessment was founded on the ability of small firms to develop markets rapidly, generate new goods and services, and offer diverse products. The study also concluded that small businesses were less risk adverse than larger, established corporations and were often better positioned to exploit market opportunities quickly. A National Science Foundation report found that entrepreneurs and small firms are six times as effective as larger firms in utilizing research and development expenditures to generate new products. National Science Board, *Science and Engineering Indicators, 1993*, (Dec. 8, 1993), 185. Anderson, Anne, "Small Businesses Make it Big in the SBIR Program," *New Technology Week* (June 6, 1998), p. 2.

[52] Sally Wyatt and Gilles Y. Bertin, *Multinationals and Industrial Property* 139 (Harvester 1988).

[53] J. Douglas Hawkins, "Importance and Access of International Patent Protection for the Independent Inventor," 3 *University of Baltimore Intellectual Property Journal* (1995), 145.

[54] P.L. 96-517, 94 Stat. 2311 (codified at 35 U.S.C. §§ 200-212).

[55] CRS Report RL32076, The Bayh-Dole Act: Selected Issues in Patent Policy and the Commercialization of Technology, by Wendy H. Schacht.

[56] Josh Lerner, "Patent Policy Innovations: A Clinical Examination," 53 *Vanderbilt Law Review* (2000), 1841.

[57] *See* Arti K. Rai and Rebecca S. Eisenberg, "Bayh-Dole Reform and the Progress of Biomedicine," 66 *Law and Contemporary Problems* (Winter/Spring 2003), 289.

[58] 35 U.S.C. § 41(g).

[59] CRS Report RS20906, U.S. Patent and Trademark Office Appropriations Process: A Brief Explanation, by Wendy H. Schacht.

[60] Under a rule of "prior user rights," when a conflict exists between an issued patent and an earlier user of the patented technology, the validity of the patent is upheld but the prior user is exempted from infringement. *See* Pierre Jean Hubert, "The Prior User Right of H.R. 400: A Careful Balancing of Competing Interests," 14 *Santa Clara Computer and High Technology Law Journal* (1998), 189. Prior user rights are discussed further in this report below.

[61] *See* Gary L. Griswold and F. Andrew Ubel, "Prior User Rights — A Necessary Part of a First-to-File System," 26 *John Marshall Law Review* (1993), p. 567.

[62] *See* David H. Hollander, Jr., "The First Inventor Defense: A Limited Prior User Right Finds Its Way Into U.S. Patent Law," 30 *American Intellectual Property Law Association Quarterly Journal* (2002), 37 (noting the perception that prior user rights favor large, well-financed corporations).

[63] In particular, economic research suggests that different industries attach widely varying values to patents. For example, one study of the aircraft and semiconductor industries suggested that lead time and the strength of the learning curve were superior to patents in capturing the value of investments. In contrast, members of the drug and chemical

industries attached a higher value to patents. Differences in the perception of the patent system have been attributed to the extent to which patents introduced significant duplication costs and times for competitors of the patentee. Richard C. Levin, Alvin K. Klevorick, Richard R. Nelson, and Sidney G. Winter, "Appropriating the Returns for Industrial Research and Development," Brookings Papers on Economic Activity, 1987, in *The Economics of Technical Change*, eds. Edwin Mansfield and Elizabeth Mansfield (Vermont, Edward Elgar Publishing Co., 1993), p. 254.

[64] Federic M. Scherer, "The Economics of Human Gene Patents", 77 *Academic Medicine* (Dec. 2002), p. 1350.

[65] Mann, *supra*, at 979.

[66] H.R. 2795, § 3.

[67] *See* Roger E. Schechter and John R. Thomas, *Principles of Patent Law* § 1.2.5 (2d ed. 2004).

[68] In addition, the party that was the first to invent must not have abandoned, suppressed or concealed the invention. 35 U.S.C. § 102(g)(2).

[69] *See* Charles E. Gholz, "First-to-File or First-to-Invent?", 82 *Journal of the Patent and Trademark Office Society* (2000), p. 891.

[70] *See* Peter A. Jackman, "Adoption of a First-to-File System: A Proposal," 26 *University of Baltimore Law Review* (1997), 67.

[71] 35 U.S.C. § 135.

[72] *See* Clifford A. Ulrich, "The Patent Systems Harmonization Act of 1992: Conformity at What Price?," 16 *New York Law School Journal of International and Comparative Law* (1996), p. 405.

[73] *See* Coe A. Bloomberg, "In Defense of the First-to-Invent Rule," 21 *American Intellectual Property Law Quarterly Journal* (1993), p. 255.

[74] *See* Bernarr A. Pravel, "Why the United States Should Adopt the First-to-File System for Patents," 22 *St. Mary's Law Journal* (1991), p. 797.

[75] 35 U.S.C. § 111(b).

[76] Gerald J. Mossinghoff, "The U.S. First-to-Invent System Has Provided No Advantage to Small Entities," 84 *Journal of the Patent and Trademark Office Society* (2002), p. 425.

[77] *See generally* Charles R.B. Marcedo, "First-to-File: Is American Adoption of the International Standard in Patent Law Worth the Price?," 18 *American Intellectual Property Law Association Quarterly Journal* (1990), p. 193.

[78] 35 U.S.C. § 102(g)(2).

[79] See Del Mar Engineering Labs. v. United States, 524 F.2d 1178 (Ct. Cl. 1975).

[80] 35 U.S.C. § 101.

[81] H.R. 2795, § 3(i).

[82] 35 U.S.C. § 102(b).

[83] Schechter and Thomas, *supra*, at § 4.3.1.

[84] European Patent Convention, Article 54(2).

[85] Japanese Patent Act, Article 29(1).

[86] 35 U.S.C. § 102(a).

[87] Convention of Paris for the Protection of Industrial Property, 13 U.S.T. 25 (1962).

[88] *See* G.H.C. Bodenhausen, *Guide to the Paris Convention for the Protection of Industrial Property* (United International Bureau for the Protection of Intellectual Property, Geneva, Switzerland 1968).
[89] 35 U.S.C. § 119(a).
[90] H.R. 2795, § 11(h).
[91] *See* Beedle v. Bennett, 122 U.S. 71 (1887).
[92] *See* Mendenhall v. Astec Indus., Inc., 13 USPQ2d 1913, 1937 (E.D. Tenn.1988), *aff'd*, 887 F.2d 1094 (Fed. Cir. 1989).
[93] *See generally* A.C. Auckerman and Co. v. R.L. Chaides.Construction Co., 960 F.2d 1020 (Fed. Cir. 1992).
[94] Schechter and Thomas, *supra*, at § 4.3.8.
[95] Bate Refrigerating Co. v. Sulzberger, 157 U.S. 1, 27 (1895) (quoting Ex parte Mushet, 1870 Comm'r Dec. 106, 108 (1870)).
[96] *See* Donald S. Chisum, "Foreign Activity: Its Effect on Patentability under United States Law," 11 *International Review of Industrial Property and Copyright Law* (1980), 26.
[97] *See* Schechter and Thomas, *supra*, at § 4.4.4.
[98] 35 U.S.C. § 111.
[99] 35 U.S.C. § 118.
[100] H.R. 2795, § 4(c).
[101] President's Commission on the Patent System,"To Promote the Progress of . . . Useful Arts" in an Age of Exploding Technology (1966).
[102] Advisory Commission on Patent Reform, *A Report to the Secretary of Commerce* (Aug. 1992), 179.
[103] Ibid.
[104] 35 U.S.C. § 112.
[105] See, e.g., Chemcast Corp. v. Arco Industries Corp. 913 F.2d 923 (Fed. Cir. 1990).
[106] *See* Jerry R. Selinger, "In Defense of the 'Best Mode': Preserving the Benefit of the Bargain for the Public, 43 *Catholic University Law Review* (1994), 1071.
[107] *See, e.g.,* Steven B. Walmsley, "Best Mode: A Plea to Repair or Sacrifice This Broken Requirement of United States Patent Law," 9 *Michigan Telecommunications and Technology Law Review* (2002), p. 125.
[108] 35 U.S.C. § 101.
[109] 1992 Advisory Commission Report, *supra,* at 102-03.
[110] *Ibid* at 101.
[111] *Ibid*. at 102-03.
[112] 35 U.S.C. § 122(a) (stating general rule that "applications for patents shall be kept in confidence by the Patent and Trademark Office and no information concerning the same given without authority of the applicant . . . .").
[113] Glaverbel Societe Anonyme v. Northlake Mktg. and Supply Inc., 45 F.3d 1550 (Fed. Cir. 1995).
[114] Heidelberger Druckmaschinen AG v. Hantscho Comm'l Prods., Inc., 21 F.3d 1068 (Fed. Cir. 1993).
[115] Jazz Photo Corp. v. U.S. Int'l Trade Comm'n, 264 F.3d 1094 (Fed. Cir. 2001).

[116] John F. Lynch, "An Argument for Eliminating the Defense of Patent Unenforceability Based on Inequitable Conduct," 16 *American Intellectual Property Law Association Quarterly Journal* (1988), 7.
[117] Burlington Indus., Inc. v. Dayco Corp., 849 F.2d 1418 (Fed. Cir. 1988).
[118] *See, e.g.,* Scott D. Anderson, "Inequitable Conduct: Persistent Problems and Recommended Resolutions," 82 *Marquette Law Review* (1999), 845.
[119] Lynch, *supra*, at 7.
[120] 1992 Advisory Commission, *supra*, at 114.
[121] H.R. 2795, § 5(a).
[122] 37 C.F.R. § 1.56.
[123] R. Carl Moy, "The Effect of New Rule 56 on the Law of Inequitable Conduct," 74 Journal of the Patent and Trademark Office Society (1992), 391.
[124] *See* Harry F. Manbeck, Jr., "The Evolution and Issue of New Rule 56," 20 *American Intellectual Property Law Association Quarterly Journal* (1992), 136.
[125] H.R. 2795, § 5
[126] Schechter and Thomas, *supra*, at § 7.2.6.
[127] John C. Todaro, "Potential Upcoming Changes in U.S. Patent Laws: the Publication of Patent Applications," 36 *IDEA: Journal of Law and Technology* (1996), 309.
[128] 35 U.S.C. § 122(b).
[129] H.R. 2795, § 9(a).
[130] 37 C.F.R. § 1.291.
[131] 35 U.S.C. § 122(c).
[132] 37 C.F.R. § 1.99.
[133] U.S. Dept. of Commerce, U.S. Patent and Trademark Off., Manual of Patent Examining Procedure § 1134.01 (8th ed. May 2004).
[134] H.R. 2795, § 10.
[135] 35 U.S.C. § 120.
[136] Schechter and Thomas, *supra*, at § 7.2.4.
[137] Mark A. Lemley and Kimberly A. Moore, "Ending Abuse of Patent Continuations," 84 Boston University Law Review (2004), 63.
[138] 35 U.S.C. § 122(b)(1).
[139] 35 U.S.C. § 122(b)(2).
[140] 35 U.S.C. § 154(a)(2).
[141] H.R. 2795, § 8.
[142] Restatement of Unfair Competition § 39.
[143] David D. Friedman, "Some Economics of Trade Secret Law," 5 *Journal of Economic Perspectives* (1991), 61, 64.
[144] 35 U.S.C. § 102(b). *See* Metallizing Engineering Co. v. Kenyon Bearing and Auto Parts, 153 F.2d 516 (2d Cir. 1946).
[145] 35 U.S.C. § 154.
[146] W.L. Gore and Associates. v. Garlock, Inc., 721 F.2d 1540 (Fed. Cir. 1983).
[147] 149 F.3d 1368 (Fed. Cir. 1998).
[148] As presently codified at 35 U.S.C. § 273, the first inventor defense is subject to a number of additional qualifications. First, the defendant must have reduced the infringing subject matter to practice at least one year before the effective filing date of the application. Second, the defendant must have commercially used the infringing

subject matter prior to the effective filing date of the patent. Finally, any reduction to practice or use must have been made in good faith, without derivation from the patentee or persons in privity with the patentee.

[149] The bill would also remove the requirement that the prior use be reduced to practice at least one year before the effective filing date of such patent. Under H.R. 2795, the defense would apply where reduction to practice occurred prior to the patent's filing date.

[150] *See* Gary L. Griswold and F. Andrew Ubel, "Prior User Rights — A Necessary Part of a First-to-File System," 26 *John Marshall Law Review* (1993), 567.

[151] Paul R. Morico, "Are Prior User Rights Consistent with Federal Patent Policy?: The U.S. Considers Legislation to Adopt Prior User Rights," 78 *Journal of the Patent and Trademark Office Society* (1996), 572.

[152] *See* Robert L. Rohrback, "Prior User Rights: Roses or Thorns?," 2 *University of Baltimore Intellectual Property Review* (1993), 1.

[153] 35 U.S.C. § 283.

[154] H.R. 2795, § 7.

[155] *See* Richardson v. Suzuki Motor Co., 868 F.2d 1226, 1247 (Fcd. Cir. 1989).

[156] MercExchange, L.L.C. v. eBay, Inc., 401 F.3d 1323, 1338 (Fed. Cir. 2005) (select citations omitted).

[157] 69 F.2d 577 (7th Cir. 1934).

[158] *Ibid.* at 593.

[159] 146 F.2d 941 (9th Cir. 1944).

[160] *Ibid.* at 943-44.

[161] 852 F. Supp. 813 (D. Minn.1994).

[162] *Ibid.* at 850 — 51.

[163] *Ibid.* at 862.

[164] 35 U.S.C. § 283.

[165] W. David Westergard, Remedying the Growing Abuse of the Patent System Through Targeted Legislation, *in Conference Papers on International Intellectual Property Law and Policy* (Fordham Law School March 31, 2005).

[166] *See* Lorraine Woellert, "A Patent War Is Breaking Out on the Hill," *BusinessWeek* 45 (July 4, 2005).

[167] Rite-Hite Corp. v. Kelley Co., Inc., 56 F.3d 1538, 1547 (Fed. Cir. 1995).

[168] *See, e.g.,* Frank H. Easterbrook, "Intellectual Property is Still Property," 13 *Harvard Journal of Law and Public Policy* (1990), 108.

[169] *See* Mark A. Lemley, "Property, Intellectual Property, and Free Riding," 83 *Texas Law Review* (2005), 1031.

[170] *See* Michael A. Carrier, "Cabining Intellectual Property Through a Property Paradigm," 54 *Duke Law Journal* (2004), 1.

[171] H.R. 2795, § 6.

[172] Westergard, *supra,* at 7.

[173] Stratoflex, Inc. v. Aeroquip Corp., 713 F.2d 1530, 1540 (Fed. Cir. 1983).

[174] Allen Eng'g Corp. v. Bartell Indus., Inc., 299 F.3d 1336, 1345 (Fed. Cir. 2002).

[175] 35 U.S.C. § 284.

[176] Ibid.

[177] *See* Read Corp. v. Portec, Inc., 970 F.2d 816, 826 (Fed. Cir. 1992).

[178] *See* Beatrice Foods Co. v. New England Printing and Lithographing Co., 923 F.2d 1576, 1578 (Fed. Cir.1991).
[179] *See, e.g.,* Jon E. Wright, "Willful Patent Infringement and Enhanced Damages — Evolution and Analysis," 10 *George Mason Law Review* (2001), 97.
[180] *See, e.g.,* Fromson v. Western Litho Plate and Supply Co., 853 F.2d 1568, 1572 (Fed. Cir. 1988).
[181] 383 F.3d 1337 (Fed. Cir. 2004).
[182] *Ibid.* at 1341.
[183] *See generally* Schechter and Thomas, *supra*, at § 9.2.5.
[184] H.R. 2795, § 6.
[185] Mark D. Janis, "Inter Partes Reexamination," 10 *Fordham Intellectual Property, Media and Entertainment Law Journal* (2000), 481.
[186] Craig Allen Nard, "Certainty, Fence Building and the Useful Arts," 74 *Indiana Law Journal* (1999), 759.
[187] *See* Schechter and Thomas, *supra*, at § 7.5.4.
[188] 35 U.S.C. § 315(c).
[189] H.R. 2795, § 9(d).
[190] *See* National Research Council of the National Academies, *A Patent System for the 21st Century* (2004), 96.
[191] *Ibid.* at 103.
[192] *See* Mark D. Janis, "Rethinking Reexamination: Toward a Viable Administrative Revocation System for U.S. Patent Law," 11 *Harvard Journal of Law and Technology* (1997), 1.

Chapter 2

PATENT LAW AND ITS APPLICATION TO THE PHARMACEUTICAL INDUSTRY: AN EXAMINATION OF THE DRUG PRICE COMPETITION AND PATENT TERM RESTORATION ACT OF 1984 ("THE HATCH-WAXMAN ACT")[*]

Wendy H. Schacht and John R. Thomas

ABSTRACT

Congressional interest in the availability of prescription drugs has focused attention on the role of patents in the pharmaceutical industry. The industry has been described as patent-intensive. Enterprises within this sector frequently obtain patent protection and enforce patent rights, and reportedly place a higher comparative value on patents than do competitors in many other markets.

The patent law is based upon the Patent Act of 1952, codified in Title 35 of the United States Code. This statute allows inventors to obtain patents on processes, machines, manufactures, and compositions of matter that are useful, new, and nonobvious. Granted patents confer the right to exclude others from making, using, selling, offering to sell, or importing into the United States the patented invention.

The Drug Price Competition and Patent Term Restoration Act of 1984 (the 1984 Act) – commonly known as the "Hatch-Waxman Act" – made several significant changes to the patent laws designed to encourage innovation in the pharmaceutical industry while facilitating the speedy introduction of lower-cost generic drugs. These changes include provisions for extending the term of a patent to reflect regulatory delays encountered in obtaining marketing approval by the Food and Drug Administration (FDA); a statutory exemption from patent infringement for activities associated with regulatory marketing approval; establishment of mechanisms to challenge the validity of a pharmaceutical patent; and a reward for disputing the validity, enforceability, or infringement of a patented and approved drug. The 1984 Act also provides the FDA with certain authorities

---

[*] Excerpted from CRS Report RL30756, dated January 10, 2005.

to offer periods of marketing exclusivity for a pharmaceutical independent of the rights conferred by patents.

Many experts agree the 1984 Act has had a significant effect on the availability of generic substitutes for brand name drugs. Lower cost generics tend to be rapidly marketed after patent expiration. Increasing investment in R&D and gains in the research intensity of the pharmaceutical industry appear to indicate that the act has not deterred the development of new drugs. However, some questioned whether the law is needed to achieve the stated goals. Critics maintained the necessity of patent-related incentives for innovation is mitigated by other federal activities. Supporters of the existing approach argued that these incentives are precisely what foster a robust pharmaceutical industry. Of fundamental interest was whether alterations of the act were in order to reflect any perceived changes in the research environment since the legislation was enacted in the 1980s.

INTRODUCTION

Congressional interest in methods to provide drugs at lower cost, particularly for the elderly, has rekindled a discussion over the role the federal government plays in facilitating the creation of new pharmaceuticals for the marketplace. Among the various federal laws that affect technology development are those dealing with intellectual property rights, particularly patents. Legislation concerning the ownership of inventions is intended to encourage additional private sector investments often necessary to further develop marketable products. The current approach attempts to balance the public sector's interest in new and improved technologies with concerns over providing companies valuable benefits without adequate accountability or compensation. Questions have been raised as to whether or not this balance is appropriate, particularly with respect to drug discovery. Critics maintain that the need for technology development incentives in the pharmaceutical and/or biotechnology sectors is mitigated by industry access to government-supported work at no cost, monopoly power through patent protection, and additional regulatory and tax advantages such as those conveyed through the Drug Price Competition and Patent Term Restoration Act and the Orphan Drug Act. Supporters of the existing approach argue that these incentives are precisely what are required and have given rise to robust pharmaceutical and biotechnology industries.

This report examines the role of patents in pharmaceutical innovation and provides an overview of the general principles of patent law as applied to inventions of the pharmaceutical industry. The study explores the provisions of several relevant statutes including the Drug Price Competition and Patent Term Restoration Act of 1984 (the 1984 Act), commonly known as the "Hatch-Waxman Act."[1] Issues and opportunities associated with the implementation of this law are addressed, since the pharmaceutical industry has been described as a patent-intensive one.[2] Enterprises within this sector frequently obtain patent protection and enforce patent rights, and reportedly place a higher comparative value on patents than do competitors in many other markets.[3]

ROLE OF PATENTS IN PHARMACEUTICAL INNOVATION

The patent system is grounded in Article I, Section 8, Clause 8 of the U.S. Constitution and is intended to stimulate new discoveries and their reduction to practice, commonly known as innovation. The Constitution states that "The Congress Shall Have Power . . . To promote the Progress of Science and useful Arts, by securing for limited Times to Authors and Inventors the exclusive Right to their respective Writings and Discoveries...." The award of a patent permits the creator of an idea to exclude others temporarily from use of that concept without compensation (currently 20 years from the date of filing). It also places the information associated with an invention within the public domain.

Patent ownership is perceived to be an incentive to innovation, the basis for the technological advancement that contributes to economic growth. It is through the commercialization and use of new products and processes that productivity gains are made and the scope and quality of goods and services are expanded. Award of a patent is intended to stimulate the investment necessary to develop an idea and bring it to the marketplace embodied in a product or process. Patent title provides the recipient with a limited-time monopoly over the application of his discovery in exchange for the public dissemination of information contained in the patent application. This is intended to permit the inventor to receive a return on the expenditure of resources leading to the discovery but does not guarantee that the patent will generate commercial benefits. The requirement for publication of the patent is expected to stimulate additional innovation and other creative means to meet similar and expanded demands in the marketplace.

Innovation typically is knowledge-driven – based on the application of knowledge, whether it is scientific, technical, experiential, or intuitive. Innovation also produces new knowledge. One characteristic of knowledge that underlies the patent system is that it is a "public good," a good that is not exhausted when it is used. As John Shoven of Stanford University points out, "[t]he use of an idea or discovery by one person does not, in most cases, reduce the availability of that information to others."[4] Therefore the marginal social cost of the widespread application of that information is near zero because the stock of knowledge is not depleted. "Ordinarily, society maximizes its welfare through not charging for the use of a free good."[5] However, innovation typically is costly and resource intensive. Patents permit novel concepts or discoveries to become "property" when reduced to practice and therefore allow for control over their use. They ". . . create incentives that maximize the difference between the value of the intellectual property that is created and used and the social cost of its creation."[6]

Studies demonstrate that the rate of return to society as a whole generated by investments in research and development (R&D) leading to innovation is significantly larger than the benefits that can be captured by the person or organization financing the work. It is estimated that the social rate of return on R&D spending is over twice that of the rate of return to the inventor.[7] Ideas often are easily imitated, the knowledge associated with an innovation dispersed and adapted to other products and processes that, in turn, stimulate growth in the economy. That can happen in the absence of appropriability defined as ". . . factors, excluding firm and market structure, that govern an innovator's ability to capture the profits generated by an innovation."[8] The appropriability of an invention depends on the level of competition in the industry and the type of information related to the innovation; the more competition and the more basic the knowledge, the less appropriable it is.[9] The difficulty in securing sufficient returns to spending on research and development has been associated with underinvestment in those activities.

The patent process is designed to resolve the problem of appropriability. If discoveries were universally available without the means for the inventor to realize a return on investments, there would result a ". . . much lower and indeed suboptimal level of innovation."[10] While research is often important to innovation, studies have shown that it constitutes only 25% of the cost of commercializing a new technology or technique. Thus, it is the expenditure of a substantial amount of additional resources that brings most products or processes to the marketplace. The grant of a patent provides the inventor with a means to capture the returns to his invention through exclusive rights on its practice for 20 years from date of filing. That is intended to encourage those investments necessary to further develop an idea and generate a marketable technology.

Issuance of a patent provides the inventor with a limited-time monopoly that is influenced by other mitigating factors, particularly the requirements for information disclosure, the length of the patent, and the scope of rights conferred. The process of obtaining a patent places the concept on which it is based in the public domain. In return for a monopoly right to the application of the knowledge generated, the inventor must publish the ideas covered in the patent. As a disclosure system, the patent can, and often does, stimulate other firms or individuals to invent "around" existing patents to provide for parallel technical developments or meet similar market needs.

The patent system thus has dual policy goals – providing incentives for inventors to invent and encouraging inventors to disclose technical information.[11] Disclosure requirements are factors in achieving a balance between current and future innovation through the patent process, as are limitations on scope, novelty mandates, and nonobviousness considerations.[12] They give rise to an environment of competitiveness with multiple sources of innovation, which is viewed by some experts as the basis for technological progress. This is important because, as Robert Merges (Boston University) and Richard Nelson (Columbia University) found in their studies, when only ". . . a few organizations controlled the development of a technology, technical advance appeared sluggish."[13]

Not everyone agrees that the patent system is a particularly effective means to stimulate innovation. It is argued that patents do not work in reality as well as in theory because they do not confer perfect appropriability. In other words, they allow the inventor to obtain a larger portion of the returns on his investment but do not permit him to capture all the benefits. Patents can be circumvented and infringement cannot always be proven. Thus, patents are not the only way, nor necessarily the most efficient means, for the inventor to protect the benefits generated by his efforts. A study by Yale University's Richard Levin and his colleagues concluded that lead time, learning curve advantages (e.g. familiarity with the science and technology under consideration), and sales/service activities were typically more important in exploiting appropriability than were patents. That was true for both products and processes. However, patents were found to be better at protecting the former than the latter. The novel ideas associated with a product often can be determined through reverse engineering – taking the item apart to assess how it was made. That information then could be used by competitors if not covered by a patent. Because it is more difficult to identify the procedures related to a process, other means of appropriation are seen as preferable to patents, with the attendant disclosure requirements.[14]

The utility of patents to companies varies among industrial sectors. Patents are perceived as critical in the drug and chemical industries. That may reflect the nature of R&D performed in these sectors, where the resulting patents are more detailed in their claims and therefore easier to defend.[15] In contrast, one study found that in the aircraft and semiconductor industries patents are not the most successful mechanism for capturing the benefits of investments. Instead, lead time and the strength of the learning curve were determined to be more important.[16] The degree to which industry perceives patents as effective has been characterized as ". . . positively correlated with the increase in duplication costs and time associated with patents."[17] In certain industries, patents significantly raise the costs incurred by nonpatent holders wishing to use the idea or invent around the patent – an estimated 40% in the pharmaceutical sector, 30% for major new chemical products, and 25% for typical chemical goods – and are thus viewed as important. However, in other industries, patents have much smaller impact on the costs associated with imitation (e.g. in the 7%-15% range for electronics), and may be considered less successful in protecting resource investments.[18]

PRINCIPLES OF PATENTABILITY

Patentable Subject Matter

Patent law is based upon the Patent Act of 1952, codified in Title 35 of the United States Code. Section 101 defines the subject matter that may be patented. According to the statute, one who "invents or discovers any new and useful process, machine, manufacture, or any composition of matter, or any new and useful improvement thereof, may obtain a patent therefore, subject to the conditions and requirements of this title."[19] An invention that falls within one of the four statutory categories – processes, machines, manufactures, and compositions of matter – may be subject to a so-called "utility patent."

Actors within the pharmaceutical industry principally claim inventions that are compositions of matter or processes. In addition to such things as mixtures and alloys, compositions of matter include chemical compounds.[20] When a composition of matter is presented in the fashion of a patent claim, it is defined in terms of its constituent elements.

A patent claim that is expressed as a series of steps is known as a process or method claim. Process claims are commonly divided into two sorts: "method of using" and "method of making" claims.[21] Suppose that an inventor manufactures a new pharmaceutical compound and also discovers that the compound may be used to treat a particular ailment. The manner in which the pharmaceutical may be employed to achieve a result may be drafted in the form of a claim towards a method of using. As well, the inventor may obtain claims for a method of making the compound, stating the techniques he employed to synthesize the compound.

Section 100(b) of the Patent Act notes that a process "includes a new use of known process, machine, manufacture, composition of matter, or method."[22] The statute thus allows inventors to obtain a proprietary interest in a newly discovered property of a known product. Suppose, for example, that an inventor discovers that a well-known chemical compound, understood to act as an explosive, also serves as a heart medication. The inventor could not obtain patent protection on a compound that already lies within the public domain. But he could seek a patent claiming a process of using the compound as a heart medication.[23]

Utility

Section 101 of the Patent Act also mandates that patents issue only to "useful" inventions.[24] Utility ordinarily presents a minimal requirement that the invention be capable of achieving a pragmatic result.[25] Patent applicants need only supply a single, operable use of the invention that is credible to persons of ordinary skill in the art. Although the utility requirement is readily met in most fields, it may present a significant obstacle to patentability for pharmaceutical inventions. Here, inventors sometimes synthesize compounds without a precise knowledge of how they may be used to achieve a practical working result. When patent applications are filed claiming such compounds, they may be rejected as lacking utility within the meaning of the patent law.

The utility requirement should be viewed in light of the considerable incentives chemists, biologists and physicians possess to obtain patent protection on compounds of interest as soon as possible. For example, in the case of pharmaceutical compounds, food and drug authorities require considerable product testing before the pharmaceutical can be broadly marketed. Before investing further time and effort on laboratory testing and clinical trials, actors in the pharmaceutical field desire to obtain patent rights on promising compounds even where their particular properties are, as yet, not well understood. But when patent applications are filed too close to the laboratory bench, inventors have discovered that the utility requirement may block the issuance of a patent.

The Supreme Court opinion in *Brenner v. Manson* addressed such a situation.[26] The inventor Manson filed a patent application claiming a method of making a known steroid compound. Although the particular compound Manson was concerned with was already known to the art, chemists had yet to identify any setting in which it could be gainfully employed. However, as skilled artisans knew that another steroid with a very similar structure had tumor-inhibiting effects in mice, Manson's new method of making the compound was a research tool of interest to the scientific community.

The U.S. Patent and Trademark Office (USPTO) Board of Patent Appeals affirmed the examiner's rejection of the application. The Board reasoned that because Manson could not identify a single use for the steroid he produced, the utility requirement was not satisfied. The Board was unimpressed that a similar compound did have beneficial effects, noting that in the unpredictable art of steroid chemistry, even minor changes in chemical structure often lead to significant and unforeseeable changes in the performance of the compound. Manson then appealed to the Court of Customs and Patent Appeals (CCPA), which reversed. Key to the CCPA's reasoning was that the sequence of process steps claimed by Manson would produce the steroid of interest. According to the CCPA, because the claimed process worked to produce a compound, the utility requirement was satisfied.

The Supreme Court granted certiorari and once more reversed, thereby upholding the Patent Office rejection. At least within the context of scientific research tools, the Court imposed a requirement that an invention may not be patentable until it has been developed to a point where "specific benefit exists in currently available form."[27] Chief among the

Court's concerns was the breadth of the proprietary interest that could result from claims such as those in Manson's application. "Until the process claim has been reduced to production of a product shown to be useful, the metes and bounds of that monopoly are not capable of precise delineation. . . . . Such a patent may confer power to block whole areas of scientific development, without compensating benefit to the public."[28] The Court closed by noting that "a patent is not a hunting license. It is not a reward for the search, but compensation for its successful conclusion. 'A patent system must be related to the world of commerce rather than to the realm of philosophy.'"[29]

Although *Brenner v. Manson* appears to take a strict view of the utility requirement, a more recent lower court opinion on utility, *In re Brana*,[30] suggests a more limited role. Like Manson, Brana claimed chemical compounds and stated they were useful as antitumor substances. The scientific community knew that structurally similar compounds had shown antitumor activity during both *in vitro* testing, done in the laboratory using tissue samples, and *in vivo* testing using mice as test subjects. The latter tests had been conducted using cell lines known to cause lymphocytic tumors in mice.

The USPTO Board rejected the application for lack of utility, and on appeal the United States Court of Appeals for the Federal Circuit (Federal Circuit) reversed. Among the objections of the USPTO was that the tests cited by Brana were conducted upon lymphomas induced in laboratory animals, rather than real diseases. The Federal Circuit responded that an inventor need not wait until an animal or human develops a disease naturally before finding a cure.[31] The USPTO further argued that Brana cited no clinical testing, and therefore had no proof of actual treatment of the disease in live animals. The Federal Circuit reasoned that proof of utility did not demand tests for the full safety and effectiveness of the compound, but only acceptable evidence of medical effects in a standard experimental animal.[32]

The holding of *Brana*, along with its failure to discuss or even cite *Brenner v. Manson*, suggests that the Federal Circuit will adopt a more liberal approach to the utility requirement than did the Supreme Court.[33] The Federal Circuit did indicate that, in cases where the invention lacks a well-established use in the art, the applicant must disclose a specific, credible use within the patent's specification.[34]

Novelty and Nonobviousness

To be patentable, a pharmaceutical invention must be judged both new and nonobvious. To be considered novel, the invention must not be wholly anticipated by the so-called "prior art," or public domain materials such as publications and other patents.[35] The nonobviousness requirement is met if the invention is beyond the ordinary abilities of a person of ordinary skill in the art in the appropriate field.[36]

Patent Acquisition Procedures

Preparing a Patent Application

An inventor who wishes to obtain patent protection must first prepare an application. Although inventors may represent themselves before the USPTO, the vast majority engage the services of a patent attorney or agent for this purpose. An application must include several components. First, the application must be accompanied by a filing fee. As of October 1, 2002, the filing fee was set to $740.[37]

The application must also contain a specification, or description of the invention. Section 112 subjects the specification to three requirements.[38] First, the specification must enable persons of ordinary skill in the art to which the patent pertains to make and use the invention.[39] Second, the specification must contain a "written description" of the invention, sufficient to show that the inventor was in possession of the invention at the time he filed the application.[40] Finally, the specification must detail the "best mode" contemplated by the inventor of practicing the invention.[41]

Section 112 also requires that the specification "conclude with one or more claims particularly pointing out and distinctly claiming the subject matter which the applicant regards as his invention."[42] The claims are considered the most important part of the patent instrument, setting forth the boundaries of the invention that the inventor claims as his own. Claims are subject to a requirement of definiteness, which mandates that they be sufficiently precise so that others may have notice of the patentee' proprietary interest.[43]

Inventors possess no duty to perform a prior art search prior to filing a patent application. However, if an applicant does know of a prior art reference that is material to the patentability of the claimed invention, then he must disclose it to the USPTO.[44]

USPTO Examination

Once an inventor has completed an application, he must forward it to the USPTO for further consideration. Prosecution of a patent at the USPTO is an *ex parte* procedure. Members of the public, and in particular the patent applicant's competitors, do not participate in patent prosecution procedures. As well, USPTO examiners do not possess a competing interest relative to the applicant. Instead, they assist the applicant in fulfilling the statutory requirements for obtaining a patent grant.[45]

Once the USPTO receives a patent application, USPTO staff will forward it to the examining group bearing responsibility for that sort of invention. A supervisory patent examiner then assigns the application to an individual examiner. The examiner will review the application and conduct a search of the prior art. The examiner then judges whether the application properly discloses and claims a patentable invention.

The examiner must notify the applicant of her response to the application. Termed an Office Action, this response may allow the application to issue or reject it in whole or in part.[46] If the claim is rejected, the examiner ordinarily must establish a prima facie case of unpatentability by a preponderance of the evidence.[47]

If a rejection has resulted, the attorney will usually respond by either amending the claims or asserting that the rejection was improper. An examiner who remains unconvinced by the applicant's response will issue a second Office Action termed a "Final Rejection." The applicant ordinarily has three options: abandon the application, persist in prosecution by filing a so-called "continuing application," or seek review of the examiner's action by filing a petition to the Commissioner or appeal to the Board of Patent Appeals and Interferences.[48]

Publication of Pending Patent Applications

The Domestic Publication of Foreign Filed Patent Applications Act of 1999 requires the USPTO to publish pending patent applications 18 months from the earliest filing date (to which they are entitled under the law).[49] Significantly, if an applicant certifies that the invention disclosed in the application will not be the subject of a patent application in another country that requires publication of applications 18 months after filing, then the application shall not be published in the United States. This act also creates provisional rights, equivalent to a reasonable royalty, owed from persons who employ the invention as claimed in the published patent application.[50]

Interferences

Sometimes multiple individuals seek patent rights on the same invention. For example, two companies may have contemporaneously developed a particular pharmaceutical and filed patent applications. In such cases, the USPTO will declare a so-called "interference" proceeding.[51] A patent interference is a complex administrative proceeding that ordinarily results in the award of a patent to one of its participants. The prevailing party in the interference is usually the individual who was the first to invent the claimed technology. [52]

Post-Grant USPTO Proceedings

USPTO involvement in the patent system does not necessarily end when it formally grants a patent. Two significant post-grant proceedings are worthy of note here. First, a patentee may employ the reissue proceeding to correct a patent that he believes to be inoperative or invalid.[53] For example, suppose that subsequent to the issuance of a patent, the patentee discovers prior art that would invalidate the patent due to anticipation or obviousnesss. By incorporating additional limitations into the patent claims through the reissue proceeding, the patentee may yet be able to define a patentable advance over the prior art.

The second significant post-grant proceeding is known as reexamination. A feature of U.S. law since 1981, the reexamination statute allows that any individual, including the patentee, a licensee, and even the USPTO Director himself, may cite a patent or printed publication to the USPTO and request that a reexamination occur.[54] If the USPTO determines that this reference raises "a substantial new question of patentability,"[55] then it will essentially reinitiate examination of the patent.[56] A certificate of cancellation results if the USPTO judges the claims to be unpatentable over the cited reference. Otherwise the USPTO issues a certificate of confirmation upholding the claims in their original or amended form.[57]

The Optional Inter Partes Reexamination Procedure Act of 1999 provides third parties with an additional option.[58] They may employ the traditional reexamination system, which has been renamed an *ex parte* reexamination. Or, they may opt for a minimal degree of participation in a newly minted *inter partes* reexamination. During *inter partes* reexamination, third party requesters may opt to submit written comments to accompany patentee responses to the USPTO. The requester may also appeal USPTO determinations that a reexamined patent is not invalid to the USPTO Board and the Court of Appeals for the Federal Circuit. To discourage abuse of *inter partes* reexamination proceedings, the statute provides that third party participants are stopped from raising issues that they raised or could have raised during reexamination.

Amendments to a patent introduced during reissue or reexamination may trigger so-called "intervening rights" that benefit competitors of the patentee. Congress recognized that third parties may have made commercial decisions based upon the precise wording of the claims of an issued patent. If these claims are later amended during reissue or reexamination, this reliance interest could be frustrated. The patent statute therefore allows the competitors of a reissued or reexamined patent to sell, continue to use, or otherwise employ the claimed invention in appropriate circumstances.[59]

Patent Term

Once the USPTO issues a patent, that patent enjoys an effective term established by the statute. The publication of this report finds the patent law in a transition period concerning patent term. For patents resulting from publications filed after June 8, 1995, the patent term is ordinarily twenty years from the date the patent application was filed.[60] For patents issued prior to June 8, 1995, as well as for patent resulting from applications pending at the USPTO as of that date, the patent endures for the greater of twenty years from filing or seventeen years from grant.[61]

Although the life of the patent is measured from the filing date, individuals gain no enforceable rights merely by filing a patent application. These rights accrue only at such time that the patent issues, and potentially include the power to enjoin infringers and obtain an award of damages. If the application was published in accordance with the Domestic Publication of Patent Applications Filed Abroad Act of 1999, then the patentee also obtains provisional rights equivalent to a reasonable royalty. Although provisional rights extend from the time the patent application was published, the patentee may not assert them until the patent issues.

Four significant qualifications may alter the basic patent term. Most significant for the pharmaceutical industry is that the term of a patent may be extended under 35 U.S.C. § 156. This provision was introduced by the Drug Price Competition and Patent Term Restoration Act of 1984.[62] This complex statute authorizes increased patent terms on inventions that have been subject to a premarket approval process under the Federal Food, Drug and Cosmetic Act.

Under 35 U.S.C. § 154(b), patentees may also obtain term extensions of up to five years due to certain prosecution delays, including the declaration of an interference of the successful pursuit of appeal to the Board of Patent Appeals and Interferences or federal court. As well, the Patent Term Guarantee Act of 1999 provides certain deadlines that, if not met by the USPTO, result in an automatic extension (day for day) of the term of individual patents. Among these deadlines are fourteen months for a First Office Action and four months for a subsequent Office Action. The prosecution also must be completed within three years of the filing date, with exceptions granted for continuing applications and appeals.

Finally, enjoyment of the full patent term is subject to the payment of maintenance fees. A patent expires after four, eight, or twelve years if maintenance fees are not timely paid on each occasion. As of October 1, 2002, the amounts due are $880 by the fourth year, $2,020 by the eighth year, and $3,100 by the twelfth year.[63]

PATENT ENFORCEMENT

The Exclusive Rights

A patent provides its proprietor with exclusive rights in the patented invention. An individual who "without authority makes, uses, offers to sell, or sells any patented invention, within the United States or imports into the United States any patented invention during the term of the patent therefor, infringes the patent."[64] Modern courts consider the phrase "patented invention" to mean the invention as recited in the claims. If an accused product or process meets every element and limitation of the claims, then the patent is said to be literally infringed.[65]

As noted, the Patent Act states that all unauthorized "uses" of the patented invention constitute an infringement. In *Roche Products, Inc. v. Bolar Pharmaceutical Co.*,[66] the Federal Circuit held that this language on its face prohibits all unauthorized uses of the patented invention, including those that might be deemed "experimental" in character. The *Roche v. Bolar* court did leave open a narrow possibility that the use of a patented invention wholly for experiment, amusement or curiosity might be judged noninfringing. However, where experimental uses of the invention are in fact motivated by commercial purposes, this "experimental use" doctrine will not serve as an infringement defense. As described above, Congress subsequently modified these "experimental use" principles with an eye towards the pharmaceutical industry.

The patent statute also includes provisions concerning contributory infringement and the active inducement of another's infringement.[67] Under these statutes, individuals who encourage the unauthorized practice of another's patent infringement may themselves be liable for patent infringement in certain circumstances. Suppose, for example, that a supplier sells a medication that has both infringing and noninfringing uses. If the supplier provides instructions, distributes advertising or offers training that promotes the infringing use, then it may be guilty of active inducement and liable for patent infringement.[68]

Although the exclusive rights provided by a patent are founded upon the claims, they are not necessarily limited to them. Although the courts have long recognized the value of clear and certain claims, they have sometimes expanded the scope of protection associated with a patent under the so-called "doctrine of equivalents." The doctrine of equivalents arose from judicial efforts to stop competitors who would introduce insignificant modifications from the claimed invention in order to avoid literal infringement.[69] As provided in the 1997 Supreme Court opinion in *Warner-Jenkinson Co. v. Hilton Davis Chemical Co.*, an accused product or process that presents insubstantial differences from the claimed invention will judged an equivalent and therefore an infringement.[70]

A defendant's intent is irrelevant to the outcome of an infringement inquiry. Even an individual who has never previously known of the asserted patent may be found to be an infringer.[71] As well, the exclusive patent rights do not provide an affirmative right for the patentee to employ the invention himself.[72] For example, the fact that an inventor obtains a patent on a pharmaceutical compound does not allow him to market this medication to others. Approval of the appropriate food and drug authorities must also be obtained.

The patents of others might also interfere with the patentee's ability to practice his own patented invention.[73] Suppose, for example, that a hypothetical entity, Alpha Co., obtains a patent on a chemical compound using for treating hypertension. Later, another hypothetical entity, Beta Co., discovers that the chemical compound is also useful for treating male pattern baldness. Even if Beta obtains a patent on a method of using the chemical to treat baldness, Beta cannot practice that method without infringing Alpha's patent. Nor can Alpha use the compound to treat baldness without infringing Beta's patent. In this case, the Alpha patent is said to be a blocking, or dominant patent over Beta's improvement or subservient patent. In such instances the holders of the dominant and subservient patent often possess incentives to cross-license one another.

The rights provided by U.S. patents are ordinarily effective only in the United States. They generally provide no protection against acts occurring in foreign countries.[74] Individuals must obtain patent protection in each nation where they wish to guard against unauthorized use of their inventions.

Under the "first sale" or "exhaustion" doctrine, an authorized, unrestricted sale of a patented product depletes the patent right with respect to that product. As a result of this doctrine, the purchaser of a patented good ordinarily may use or resell the good without further regard to the patentee. The courts have reasoned that when a patentee sells a product without restriction, it impliedly promises its customer that it will not interfere with the full enjoyment of the product.[75]

The Process Patents Amendment Act of 1988

Special infringement provisions concerning process patents impact the pharmaceutical industry. Traditionally the patent law held that a process claim could be directly infringed only by the performance of those steps. Suppose, for example, that an inventor holds a patent on a particular method of making a pharmaceutical. By itself, the act of selling the pharmaceutical does not infringe this method patent. The seller would also have to make the pharmaceutical by the patented method in order to be liable for infringement.[76]

This general principle was altered to some degree in the Process Patents Amendment Act of 1988.[77] There, Congress provided process patent owners with the right to exclude others from using or selling in the United States, or importing into the United States, products made by a patented process.[78] For example, suppose that an enterprise based abroad manufactures a pharmaceutical employing a process patented in the United States. If the foreign company exports the pharmaceutical into the United States, it may face liability even though it performed every step of the patented process abroad.

A number of exceptions limit liability under the Process Patents Amendment Act. If the accused product is materially changed by subsequent processes, or becomes a trivial or nonessential component of another product, then there is no infringement.[79] The Process Patents Amendment Act also included complex provisions that modified the usual scheme of remedies available for patent infringement.[80] Among other features, they include a grace period for individuals unaware of the patent implications of a particular process. Such persons may, upon receiving notice of infringement, dispose of infringing products and avoid liability.

The Process Patents Amendment Act also modified the burden of proof for certain charges of process patent infringement. Ordinarily, the patentee is the moving party during infringement litigation and bears the burden or proving that infringing acts have occurred.[81] However, Congress recognized that patentees may face great difficulties in proving that a particular product resulted from the performance of the patented process. The Patent Act therefore creates a presumption that a product is made by a patented process if two conditions are met.[82] First, there must be a substantial likelihood that the product was made by the patented process. Second, the plaintiff must have made a reasonable effort to determine the process actually used in the production of the product and was unable to so determine. The effect of the presumption is that the accused infringer has the burden of asserting that the accused product was not made by the patented process.

Infringement Litigation

The patentee may file a civil suit in federal district court in order to enjoin infringers and obtain monetary remedies.[83] Although issued patents enjoy a presumption of validity, accused infringers may assert that the patent is invalid or unenforceable.[84] In patent matters, appeals from the district courts go to the United States Court of Appeals for the Federal Circuit. The Federal Circuit also hears appeals from the USPTO. Federal Circuit decisions are subject to review at the Supreme Court.[85]

Remedies

The Patent Act sets forth the remedies a patentee may obtain upon a finding of infringement. Section 283 allows courts to "grant injunctions in accordance with the principles of equity to prevent the violation of any right secured by patent, or such terms as the court deems reasonable."[86] A patentee may also obtain a preliminary injunction against an accused infringer. Courts assess the traditional four factors when considering whether to grant such an injunction. The factors are typically stated as: (1) the probability of success on the merits; (2) the possibility of irreparable harm to the patentee if the injunction is not granted; (3) the balance of hardships between the parties; and (4) the public interest.[87]

The Patent Act also provides for the award of damages "adequate to compensate for the infringement, but in no event less than a reasonable royalty for the use made of the invention by the infringer."[88] In practice, patentees seek lost profits damages when they are able to make the required showing. Otherwise a reasonable royalty serves as the default measure of damages. The Patent Act limits recovery to six years prior to the filing of the complaint or counterclaim for patent infringement.[89] Courts ordinarily award prejudgment interest in order to afford the patentee full compensation for the infringement.[90]

PATENT ASSIGNMENTS AND LICENSES

Patents possess the attributes of personal property and may be assigned or licensed to others.[91] An assignment, which is essentially the sale of the patent, must be in writing to be effective.[92]

A patent owner may also grant a license. A license is generally not a full ownership interest in the patented invention. Instead, a patent license amounts to a promise by the patentee not to sue the licensee for infringement in exchange for some consideration.[93] Licenses are generally classified as either exclusive or nonexclusive. An exclusive licensee has received a promise that it alone may make, use, sell, offer to sell, or import into the United States the patented invention without facing an infringement suit.[94]

THE DRUG PRICE COMPETITION AND PATENT TERM RESTORATION ACT OF 1984

The Drug Price Competition and Patent Term Restoration Act of 1984 (the 1984 Act)[95] introduced several significant changes to the patent laws. These include patent term extension; a statutory exemption for patent infringement relating to regulatory marketing approval; procedures for challenging the validity of pharmaceutical patents; and a reward for challenging the validity, enforceability, or infringement of a patented and approved drug. Through these provisions, the 1984 Act attempts to balance two competing objectives within the pharmaceutical industry. First, the 1984 Act aimed to encourage the introduction of widely available generic drugs. Second, the 1984 Act hoped to ensure that adequate incentives remain for individuals to invest in the development of new drugs.[96]

The 1984 Act is today commonly known as the "Hatch-Waxman Act."[97] At the time of its enactment, however, the 1984 Act was generally referred to as the "Waxman-Hatch Act."[98] In light of this conflicting nomenclature, this report refers to the Drug Price Competition and Patent Term Restoration Act of 1984 as the 1984 Act.

Background of the 1984 Act

The Role of the FDA and the USPTO in the Pharmaceutical Industry

Both the Patent and Trademark Office and the Food and Drug Administration (FDA) have a role to play in the pharmaceutical industry. The USPTO allows patents to issue on the compounds that comprise a pharmaceutical as well as methods of making and using them. Patents confer the right to exclude others from making, using, selling, offering to sell, or importing into the United States the patented invention.[99]

The grant of a patent does not provide its proprietor with the affirmative right to market the patented invention, however.[100] For many products of the pharmaceutical industry, the FDA must approve the product for sale to consumers. Federal laws generally require that pharmaceutical manufacturers show their products are safe and effective in order to market these products.[101]

USPTO issuance of a patent and FDA marketing approval are distinct events that depend upon different criteria.[102] The FDA might consider a pharmaceutical safe and effective for consumer use, for example, but the USPTO could rule that the compound does not present a sufficient advance over public domain knowledge to be worthy of a patent. Alternatively, it is readily within the power of the FDA to judge that a pharmaceutical presents too great a risk for use as a medication within the United States, despite the fact that the USPTO has allowed a patent to issue claiming that pharmaceutical.

As a result of the independence of patent ownership and marketing approval, the pharmaceutical industry must account for both. In order to sell a drug without fear of civil or criminal liability, an enterprise must both obtain FDA approval and consider whether that drug has been patented. Often the entity which owns the patent on a pharmaceutical is the first to be awarded marketing approval. Sometimes the enterprise which has been awarded marketing approval and the patent owner are separate entities, however. In this latter case, the patentee may commence infringement litigation against the approved drug manufacturer. A court may issue an injunction and award monetary liability for patent infringement despite the fact of FDA marketing approval.

Although the 1984 Act maintained the independence between the award of a patent and the process of seeking FDA market approval, it did establish a procedural interface between these two events. Before describing these procedures in greater detail, this report first considers core features of the patent and food and drug laws as they stood prior to the 1984 Act.

The Generic Drug Approval Process

Since 1962, federal law has required pharmaceutical manufacturers to demonstrate that their products are safe and effective.[103] Prior to the 1984 Act, however, the federal food and drug law contained no separate provisions addressing generic versions of drugs that had previously been approved.[104] The result was that would-be generic drug manufacturers had to file their own "New Drug Application" (NDA) in order to market their drug. Some generic manufacturers could rely on published scientific literature demonstrating the safety and efficacy of the drug. These sorts of studies were not available for all drugs, however. Further, at times the Food and Drug Administration requested additional studies to deal with safety and efficacy questions that arose from experience with the drug following its initial approval. The result is that some generic manufacturers were forced to prove independently that the drug was safe and effective, even though their product was identical to that of a previously approved drug.

Some commentators believed that the approval of a generic drug was a needlessly costly, duplicative and time-consuming process prior to the 1984 Act.[105] FDA safety and efficacy requirements sometimes required clinical trials, for example, which could prove very expensive. Some observers noted that although patents on important drugs had expired, manufacturers were not moving to introduce generic equivalents for these products.[106] As the introduction of generic equivalents often causes prices to decrease, the interest of consumers was arguably not being served through these observed costs and delays.[107]

Generic Drug Development and Patent Infringement

The patent law grants patent proprietors the right to exclude others from making, using selling, offering to sell, or importing into the United States the patented invention.[108] Accused infringers may offer several defenses to avoid liability for patent infringement, however. One potential defense lies under the so-called "experimental use" doctrine. Perhaps the first discussion of this infringement defense occurred in the 1813 decision in *Whittemore v. Cutter*.[109] There, Justice Joseph Story explained that "it could never have been the intention of the legislature to punish a man, who constructed such a [patented] machine merely for philosophical experiments, or for the purpose of ascertaining the sufficiency of the machine to produce its described effects." By 1861, the court in *Poppenhausen v. Falke* was able to state that the law was "well-settled that an experiment with a patented article for the sole purpose of gratifying a philosophical taste, or curiosity, or for mere amusement is not an infringement of the rights of the patentee."[110]

Commentators have noted that the number of accused infringers who have successfully pled an experimental use defense are few, however.[111] As a practical matter, perhaps infringement charges were only rarely brought against philosophers or amusement seekers.[112] The possibility of an experimental use defense took on a new characteristic with the advent of drug marketing approval procedures, however. When a competitor becomes interested in marketing the generic equivalent of a drug patented by another, it may wish to commence the clinical trials and other procedures during the term of the patent. As a result, the competitor would be able to market the drug immediately upon expiration of the patent. Whether the regulatory compliance activities of a generic drug manufacturer amounted to a patent infringement, or were exempted by the experimental use defense, was for many years an open legal question.

The 1984 decision of the Court of Appeals for the Federal Circuit in *Roche Products, Inc. v. Bolar Pharmaceutical Co.*[113] resolved this question conclusively in favor of a finding of patent infringement. In that case, Roche Products, Inc. (Roche) marketed a prescription sleeping pill under the trademark "Dalmane." Roche also was the proprietor of a patent claiming a chemical compound, flurazepam hcl, that was the active ingredient in Dalmane.[114] The Roche patent issued on January 17, 1967, and expired on January 17, 1984.

Bolar Pharmaceutical Co. (Bolar), a manufacturer of generic drugs, grew interested in marketing a generic equivalent of Dalmane. Bolar recognized that FDA approval of a drug was a time-consuming process and wished to begin selling a generic equivalent immediately after the Roche patent expired. As a result, in mid-1983, Bolar obtained a supply of flurazepam hcl from a foreign manufacturer. It began to form the flurazepam hcl into dosage form capsules to obtain stability data, dissolution rates, bioequivalency studies and blood serum studies necessary to file an NDA with the FDA.

Roche brought suit against Bolar on July 28, 1983, seeking to enjoin Bolar from using flurazepam hcl for any purpose during the life of the patent. The district court ultimately denied Roche's request on October 11, 1983. The district court concluded that Bolar's use of the compound for federally mandated testing did not infringe the Roche patent because Bolar's use was minimal and experimental.[115]

Roche promptly appealed to the United States Court of Appeals for the Federal Circuit, which reversed the district court. Writing for a three-judge panel, Judge Nichols initially observed that the 1952 Patent Act states that whoever "uses . . . any patented invention, within the United States during the term of the patent therefore, infringes the patent."[116] This language on its face prohibits all unauthorized uses of the patented invention, the Federal Circuit reasoned, and many judicial opinions had so held.[117]

The Federal Circuit next considered two contentions offered by Bolar. First, Bolar urged that the experimental use defense exempted its efforts to comply with federal food and drug law. After reviewing the precedents, Judge Nichols disagreed, concluding:

> Bolar's intended "experimental" use is solely for business reasons and not for amusement, to satisfy idle curiosity, or for strictly philosophical inquiry. Bolar's intended use of flurazepam hcl to derive FDA required test data is thus an infringement of the [Roche] patent. Bolar may intend to perform "experiments," but unlicensed experiments conducted with a view to the adaptation of the patented invention to the experimentor's business is a violation of the rights of the patentee to exclude others from using his patented invention. It is obvious here that it is a misnomer to call the intended use de minimus. It is no trifle in its economic effect on the parties even if the quantity used is small. It is not dilettante affair such as Justice Story envisioned. We cannot construe the experimental use rule so broadly as to allow a violation of the patent laws in the guise of "scientific inquiry," when that inquiry has definite, cognizable, and not insubstantial commercial purposes.[118]

Bolar finally urged the Federal Circuit to resolve a perceived conflict between the Food, Drug and Cosmetic Act[119] and the 1952 Patent Act.[120] Bolar observed that substantial regulatory delays were associated with the receipt of FDA marketing approval. According to Bolar, if a generic manufacturer could not commence seeking FDA approval until the appropriate patents had expired, then the patentee could preserve its market exclusivity beyond the statutory patent term. Bolar characterized this situation as a de facto patent term extension inconsistent with the Patent Act.[121]

The Federal Circuit also rejected this argument. According to Judge Nichols, the judiciary was not the proper forum to engage in policy argumentation inconsistent with the patent statute. The court observed that bills addressing these issues had been placed before Congress and suggested that any aggrieved parties seek redress there.[122] The Federal Circuit remanded the decision to the district court with instructions to fashion the appropriate remedy.[123]

Principal Provisions of the 1984 Act

The Federal Circuit's suggestion that a legislative forum may better suit the interests of the parties proved prophetic. On September 24, 1984, President Ronald Reagan signed into law the Drug Price Competition and Patent Term Restoration Act of 1984 ("the Hatch-Waxman Act"). The 1984 Act is codified in Titles 15, 21, 28 and 35 of the United States Code.[124] Although the 1984 Act is a complex statute, observers have frequently noted that it presents a fundamental trade-off: In exchange for permitting manufacturers of generic drugs to gain FDA marketing approval by relying on safety and efficacy data from the original manufacturer's NDA, the original manufacturers received a period of data exclusivity and patent term extension.[125] A review of the legislation's more significant provisions follows.

Accelerated Generic Drug Approval Process

The 1984 Act created a new type of application for market approval of a pharmaceutical. This application, termed an Abbreviated New Drug Application (ANDA), may be filed at the FDA.[126] An ANDA may be filed if the active ingredient of the generic drug is the bioequivalent of the approved drug. An ANDA allows a generic drug manufacturer to rely upon the safety and efficacy data of the original manufacturer. The availability of an ANDA often allows a generic manufacturer to avoid the costs and delays associated with filing a full-fledged NDA. Through the ANDA procedure, a generic manufacturer may often place its FDA-approved bioequivalent drug on the market as soon as the patent on the original drug expires.[127]

Patent Term Restoration

The 1984 Act also provides for the extension of patent term. Ordinarily, patent term is set to twenty years from the date the patent application is filed.[128] The 1984 Act provides that for pharmaceutical patents, the patent term may be extended for a portion of the time lost during clinical testing. More specifically, this term extension is equal to the time between the effective date of the investigational new drug application and the submission of the NDA, plus the entire time lost during FDA approval of the NDA.[129]

The 1984 Act sets some caps on the length of the term restoration. The entire patent term restored may not exceed five years. Further, the remaining term of the restored patent following FDA approval of the NDA may not exceed 14 years.[130] The 1984 Act also provides that the patentee must exercise due diligence to seek patent term restoration from the USPTO, or the period of lack of diligence will be offset from the augmented patent term.[131]

Patent term extension does not occur automatically. The patent owner or its agent must file an application with the USPTO requesting term extension within 60 days of obtaining FDA marketing approval. According to a senior legal advisor in the Special Program Law Office of the Patent and Trademark Office, between 50 and 60 such applications are filed each year.[132]

Market Exclusivity

The 1984 Act includes provisions that create market exclusivity for certain FDA-approved drugs. The FDA administers these provisions by issuing approval to market a pharmaceutical to only a single entity. A grant of market exclusivity does not depend on the existence of patent protection and the two rights may actually conflict.

The length of market exclusivity is contingent on whether or not the drug is considered a new chemical entity (NCE). The 1984 Act defines an NCE drug as an approved drug which consists of active ingredients, including the ester or salt of an active ingredient, none of which has been approved in any other full NDA.[133] If the approved drug is not an NCE, then the FDA may not approve an ANDA for a generic version of the approved drug until three years after the approval date of the pioneer NDA.[134]

In contrast, if the approved drug is an NCE, then a would-be generic manufacturer cannot submit an ANDA until five years after the date of the approval of the pioneer NDA.[135] The effect of this provision is to restrict a potential generic manufacturer from bringing a product to market for five years plus the length of the FDA review of the ANDA. One noted expert has recently observed that the review time for an ANDA exceeds 18 months.[136]

Patent Infringement

The 1984 Act includes elaborate provisions governing the mechanisms through which a potential generic manufacturer may obtain market approval on a drug that has been patented by another. Among these provisions are a statutory exemption from claims of patent infringement based on acts reasonably related to seeking FDA approval; special provisions for challenging the enforceability, validity or infringement of approved drug patents; and a reward for challenging patent enforceability, validity or infringement consisting of 180 days of market exclusivity to the first generic applicant to file a patent challenge against any approved drug.

The 1984 Act modified the 1952 Patent Act by creating a statutory exemption from certain claims of patent infringement. As codified in § 271(e)(1), this provision mandates that "It shall not be an infringement to make, use, offer to sell, or sell within the United States a patented invention . . . solely for uses reasonably related to the development and submission of information under a Federal Law which regulates the manufacture, use or sale of drugs or veterinary biological products." This provision effectively overturns the opinion of the Court of Appeals for the Federal Circuit in *Roche Products, Inc. v. Bolar Pharmaceutical Co., Inc.*[137] As a result, generic manufacturers may commence work on a generic version of an approved drug any time during the life of the patent, so long as that work furthers compliance with FDA regulations.

Courts have interpreted § 271(e)(1) liberally, reasoning that the statute exempts from infringement a wide variety of acts. Exemplary is the decision of United States Magistrate Judge Brazil in *Intermedics, Inc. v. Ventritex, Inc.*[138] There, the court reasoned that it would not always be clear to prospective pharmaceutical suppliers exactly which kinds of information, and in what quantities, would be required to obtain FDA approval. The court therefore concluded that parties should be given some latitude in making judgments about the nature and extent of otherwise infringing activities needed to generate information that would satisfy the FDA.

The *Intermedics* court then applied this reasoning to the facts before it, concluding that a number of accused activities fell within the safe harbor of § 271(e)(1). The court held that device sales to foreign distributors were reasonably related to developing information to be submitted to the FDA because all of the devices were resold to FDA-approved clinical investigators.[139] Foreign testing activities were also found noninfringing because the data they generated was also sent to the FDA.[140]

The Supreme Court decision in *Eli Lilly and Co. v. Medtronic* is also notable for its expansive interpretation of § 271(e)(1).[141] There, the Court held that the infringement exemption is available not only to drug and veterinary products, but also to medical devices that cannot be marketed without Food and Drug Administration approval.

Although the 1984 Act provides a safe harbor from patent infringement, it also requires would-be manufacturers of generic drugs to engage in a specialized certification procedure. The core feature of this process is that a request for FDA marketing approval is treated as an "artificial" act of patent infringement. This feature was intended to allow judicial resolution of the validity, enforceability and infringement of patent rights before generic competition enters the market.[142]

Under the 1984 Act, each holder of an approved NDA must list pertinent patents it believes would be infringed if a generic drug were marketed before the expiration of these patents. The FDA publishes this list of patents in its list of approved products.[143] This list is commonly known as the "Orange Book."[144]

An ANDA applicant must certify its intent with regard to each patent associated with the generic drug it seeks to market. Four possibilities exist under the 1984 Act:

- that patent information on the drug has not been filed;
- that the patent has already expired;
- the date on which the patent will expire; or
- that the patent is invalid or will not be infringed by the manufacture, use or sale of the drug for which the ANDA is submitted.

These certifications are respectively termed paragraph I, II, III, and IV certifications.[145] An ANDA certified under paragraphs I or II is approved immediately after meeting all applicable regulatory and scientific requirements.[146] An ANDA certified under paragraph III must, even after meeting pertinent regulatory and scientific requirements, wait for approval until the drug's listed patent expires.

If the ANDA applicant files a paragraph IV certification, it must notify the proprietor of the patent. The patent owner may bring a patent infringement suit within 45 days of receiving such notification.[147] If the patent owner timely brings a patent infringement charge against the ANDA applicant, then the FDA must suspend approval of the ANDA until one of the following events occurs:

- the date of the court's decision that the listed drug's patent is either invalid or not infringed;
- the date the listed drug's patent expires, if the court finds the listed drug's patent infringed;[148] or
- subject to modification by the court, the date that is thirty months from the date the owner of the listed drug's patent received notice of the filing of a Paragraph IV certification.[149]

The 1984 Act provides prospective manufacturers of generic pharmaceuticals with a reward for challenging the patent associated with an approved pharmaceutical. The reward consists of a 180-day generic drug exclusivity period awarded to the first generic applicant to file a paragraph IV certification. This provision is intended to encourage generic applicants to challenge a listed patent for an approved drug product.[150]

The decision of the United States Court of Appeals for the D.C. Circuit in *Mova Pharmaceutical Corp. v. Shalala* considered the 180-day exclusivity provision and its implementation by the FDA.[151] Before *Mova*, the FDA took the position that in order to win the 180-day exclusivity period, the generic applicant had to defend successfully a patent infringement suit brought by the patentee under paragraph IV. In *Mova*, the D.C. Circuit held that the FDA had improperly imposed this requirement of a successful defense. According to

Judge Wald, this requirement was "gravely inconsistent with the text and structure of the statute."[152]

The holding in *Mova* may be considered in light of the reality that no provision of the 1984 Act requires the first entity to challenge a patent to pursue that challenge diligently in the courts. The first patent opponent may file a paragraph IV certification, be charged with infringement by the patentee, and then simply decide not to pursue the matter further. Nonetheless, if the patent has not yet expired, the 1984 Act prevents the FDA from approving a subsequently filed ANDA until 180 days after either (a) a court holds the challenged patent invalid, not infringed or unenforceable; or (b) the first patent challenger markets the pertinent pharmaceutical.[153]

Suppose, for example, that generic manufacturer "Alpha" is the first to file a paragraph IV certification. The patentee then commences patent infringement litigation against Alpha in the courts. Assume further that Alpha loses, or that Alpha has a change of heart and decides not to further contest the charge of infringement. Another generic manufacturer, "Beta," then files its own paragraph IV certification. Following a patent infringement lawsuit brought by the patentee against Beta, the courts hold that the patent was invalid.

Under these circumstances, the FDA may not approve a subsequently filed ANDA until Beta has obtained a judicial judgment adverse to the patent. Further, the FDA must wait 180 days after the court's judgment before granting market approval to Beta. Because Beta was not the first to challenge the patent, Beta receives no market exclusivity under the 1984 Act.

Subsequent Legislative Developments

Two significant legislative developments occurred subsequent to the enactment of the 1984 Act. First, Congress incorporated animal drugs into the structure of the 1984 Act with the 1988 Generic Animal Drug and Patent Term Restoration Act.[154]

Second, the Uruguay Round Agreement Act (URAA),[155] also amended the 1984 Act. Among the provisions of the URAA were changes to the term for which patents endure. Prior to the URAA, patents expired 17 years after the date they issued. The URAA provided that patent term would be set to 20 years from the date the patent application was filed. The URAA also included a transitional provision: patents in effect on June 8, 1995, or patent applications pending at the USPTO on that date would get the term of 20 years from the filing date or 17 years from the issue date, whichever was longer. Because the USPTO had issued many patents less than three years after an application had been filed, this so-called "Delta Period" amounted to a patent term extension.[156]

The drafters of the URAA recognized that some individuals may have made commercial plans based on the date they believed a competitor's patent would expire. Such plans would be upset if the term of the patent was unexpectedly increased. The URAA therefore included provisions that accounted for the interests of the patentee's competitors. In essence, the URAA denied the patentee the ability to prevent competitors from using the patented invention during the Delta Period. Instead, the patentee may claim an "equitable remuneration" from those who use the patented invention during the Delta Period. These provisions in effect call for a compulsory license.[157]

Although they are not formally associated with the 1984 Act, legislation relating to orphan and pediatric drugs is worthy of mention here. Both the Orphan Drug Act[158] and the Food and Drug Administration Modernization Act,[159] as amended by P.L. 107-109, the Best Pharmaceuticals for Children Act, encourage the research, development and marketing of certain drugs. The Orphan Drug Act provides drug researchers and manufacturers with several incentives concerning pharmaceuticals effective against rare diseases or conditions. These include federal funding of grants and contracts for clinical trials of orphan products; a tax credit of fifty percent of clinical testing costs; and the grant of an exclusive right to market the orphan drug for seven years from the date of FDA marketing approval.[160]

The Food and Drug Modernization Act aimed to increase the number of pharmaceuticals available for children.[161] The act provides a so-called "pediatric exclusivity" to encourage drug manufacturers to conduct research concerning the effectiveness of their drugs in children. Pediatric exclusivity attaches to any children's drug products with the same so-called "active moiety," which is that portion of the drug that causes its physiological or pharmacological reaction.[162] It typically extends the approved manufacturer's existing protection for an additional six months.[163] The product must be one for which studies on a pediatric population are submitted at the request of the Secretary of Health and Human Services. Note that the Food and Drug Administration Modernization Act does not require that a study be successful in demonstrating safety and effectiveness in a pediatric population in order to trigger the added six-month exclusivity period. Thus, the statute is merely intended to create incentives for enterprises to conduct research and submit their results.[164]

IMPLEMENTATION OF THE 1984 ACT

There has been on-going congressional interest in the 1984 Act since it was passed 18 years ago. Current concerns over the price and availability of drugs in the United States has again focused attention on the legislation because of its effort to balance innovation in the pharmaceutical industry and costs to the public. In attempting to determine any results of the implementation of the 1984 Act, it is necessary to consider the state of the pharmaceutical industry in order to assess changes in both the generic drug and brand name (or innovator) drug markets. The relationship between these sectors was the basis for prior congressional action; whether and/or how this relationship has changed to meet the objectives of the law underlies any future discussion on the 1984 Act.

Brief Overview of the Pharmaceutical Industry

The U.S. pharmaceutical industry is "highly innovative and technologically advanced . . . [and] has consistently maintained a competitive edge in international markets."[165] According to the U.S. Department of Commerce, the industry is expected to experience continued growth.[166] Much of this is the result of the substantial investment in research and development. Information provided by the National Science Foundation indicates that R&D performance in the United States by the pharmaceutical industry rose from $1.8 billion in 1980 to $6.3 billion in 1990 and $12.2 billion in 1999.[167] According to the Pharmaceutical Research and Manufacturers of American (PhRMA), U.S. R&D expenditures (domestic and foreign firms) have increased substantially during the period under consideration here: from

$1.6 billion in 1980 to $6.8 billion in 1990, to $17.2 billion in 1998, to $25.7 billion in 2002.[168] As a result of this investment, approximately 1,000 new pharmaceuticals are currently in the process of being brought to the marketplace.

Concurrently, federally-funded research is playing a significant role in private sector R&D, including in the pharmaceutical industry. In FY2000, the National Institutes of Health (NIH) supported $15.7 billion in health-related R&D. This figure represents approximately 20% of the total federal R&D budget, second only to the research funding spent for defense. According to the last relevant survey conducted by NIH, in FY1995 the federal government provided 37% of the total national support for health R&D or $13.4 billion, industry supplied 52% or $18.6 billion, and private non-profits (4% or $1.3 billion), as well as state and local government (7% or $2.4 billion), funded the remainder.[169] These figures show a change from ten years earlier when the federal government provided 46% of national health-related R&D, while industry funded 42% of the total amount spent.

During the 1980s and 1990s, the pharmaceutical industry was among the most profitable of industrial sectors based on standard accounting principles for rate of return. However, these rates are somewhat lower if additional (and significant) investments in research and advertising are accounted for.[170] While profitable, this industry also has become increasingly research intensive, reinvesting sizeable portions of profits back into R&D.[171] The ratio of R&D investment to total sales in the pharmaceutical industry has increased from 8.9% in 1980 to 16.1% in 2002. This compares to an average 4% R&D-to-sales ratio for all U.S. industries.[172]

Effects on Generic Drugs

Many experts agree that the Drug Price Competition and Patent Term Restoration Act has had a significant effect on the availability of generic substitutes for brand name drugs. "As a result of the 1984 Act, generic firms now enter the market much more rapidly after patent expiration and enter in abundant numbers."[173] Prior to the law, 35% of top-selling drugs had generic competitors after patent expiration; now almost all do.[174] In addition, the time to market for these generic products has decreased substantially. According to the Congressional Budget Office (CBO), the average time between the expiration of a brand name patent and the availability of a generic was three years before passage of the 1984 Act. Currently, the generic may be introduced immediately after the original patent expiration if it has received the approval of the FDA as companies are permitted to undertake clinical testing during the time period a patent is in force. In cases where the generic manufacturer is the patent holder, a substitute drug may be brought to market before the patent expires.

The number of prescriptions filled by generics has increased. In 1980, 69% of the prescriptions that were filled in the United States were for drugs that had multiple sources; yet, even in those cases where several drugs were available, generics were substituted in only 25% of the applicable situations.[175] Research conducted by Sherer indicated that the rate generics were dispensed in retail pharmacies rose from 17% in 1980 to 30% in 1989.[176] Similarly, CBO found that in 1980 13% of the prescriptions for multi-source drugs were filled by generic prescriptions; by 1998 they comprised 58% of the total. According to PhRMA, the generic share of the prescription drug market (measured in countable units such as tablets) rose from 18.6% in 1984 when the legislation was passed to 47% in 2000.[177] Almost identical figures are provided by the Generic Pharmaceutical Association (GPhA), the difference being a slightly lower 44% market share for generics in 2000.[178]

Reflecting the lower cost of generic drugs, these drugs represent a much smaller percent of total pharmaceutical sales dollars; 8.4% as compared with brand name drugs at 91.6% of the total spent. GphA estimates that U.S. retail sales of generics totaled $11.1 billion in 2001.[179] Prices for generic drugs tend to fall over time.[180] It should be noted, however, that the market share of generic drugs is not just dependent on prices; other factors such as perception of quality, as well as first to market, also make a difference.[181]

Effects on Brand Name Drugs

While the 1984 Act has led to a discernable increase in the availability of generic drugs, the effects on brand name pharmaceuticals appears more complex. The data suggests that R&D funding, as well as R&D intensity, are increasing. While there are no direct measures of innovation, these figures, along with the number of new drugs approved and those in development, do provide indicators of continuing innovation in the industry. However, it is not clear whether the innovation occurring is facilitated by the 1984 Act or is independent of its provisions. Some experts argue that the expiration of patents and the desire to generate new replacement drugs, not the extension of patent ownership, is the stimulus to innovation.[182]

The portions of the legislation that have accelerated the introduction of generic products have affected the brand name firms in various ways that may or may not influence innovation in the industry. The Congressional Budget Office found that originator drugs lose more than 40% of their market, on average, to generic versions after a patent expires. This is combined with research that indicates the rate of market share decline is increasing. Studies by Grabowski and his colleagues indicate that while these brand name drugs lost more than 31% of their market share (per unit) in the year between 1989 and 1990, during the first six months of 1993, 50% of market share was lost. The larger "blockbuster" drugs lost up to 90% of sale revenue within one year of the expiration of the patent.[183]

Despite competition from generics that have appreciably lower prices, the prices for brand name drugs often increase after patent expiration. Grabowski and Vernon found that innovator drug prices continued to increase at the same rate as before the introduction of generics even as market shares declined. At the same time, generic prices for the comparable drugs fell.[184] Brand name firms have reacted to the opportunities for establishing a generic market provided in the 1984 Act by "maintaining and even raising the price of the brand-name product on the theory that the demand for it was more inelastic than the demand for the price-sensitive segment; they have embarked on a new aggressive strategy designed to serve the brand-loyal segment and capture a substantial share of the generic market."[185]

Such price increases are based on the recognition that when generic substitutes are available, the market bifurcates. Price-insensitive consumers will pay more for a brand name while consumers that respond to price will buy the generic.[186] One expert, F.M. Scherer notes that ". . .price competition worked much more powerfully *among* relatively undifferentiated generic products than *between* differentiated branded products and undifferentiated generics."[187] To protect their market share, brand name companies focus on developing brand loyalty. They also may encourage doctors to move to improved versions of the drug still covered by patents.[188] An indication of what might be considered the success of this approach is contained in the observation that innovator drugs ". . .keep about half their market in units despite the fact that generics are roughly one-third the price of pioneers [innovator drugs] (measured two years after entry)."[189]

The 1984 Act created mechanisms to address concerns that regulatory requirements for FDA approval of a drug prior to marketing often meant that the owner of a patent associated with a drug did not enjoy the full benefit conferred by that patent. Provisions were included to extend the patent as compensation for some of the regulatory activities.[190] As a result, many experts have concluded that the average effective patent life today is slightly longer than before passage of the 1984 Act. According to CBO, prior to the implementation of this legislation, the average effective patent life of a pharmaceutical was approximately nine years. Today it is approximately 11.5 years. Research performed by Grabowski and Vernon and reported in 1996 indicates that for the period of time between 1991 and 1993, the 1984 Act ". . .has led to modest increases in patent terms."[191] During these years the average patent life for new drug introductions was 11.7 years, including an average extension of 2.3 years. The maximum five year extension was provided to 9% of the new drug introductions and 34% obtained an extension of over three years. Other industries average 18 years of effective patent life.[192]

A study by the University of Minnesota's Institute of Pharmaceutical Research in Management and Economics (and funded in part by generic drug manufacturers) looked at the range of patent protection of several major drugs.[193] The researchers at the University found the following:

Drug	Company	Current Patent Protection
Claritin	Schering Plough	9.2 years
Relafen	SmithKline Beecham	11 years
Cardiogen-82	Bristol-Meyers Squibb	12.7 years
Eulexin	Schering Plough	12.3 years
Nimotop	Bayer	13.8 years
Dermatop	Hoechst Marion Roussel	6.8 years
Penetre	Rhone-Poulenc Rorer	9.9 years

In addition to, and separate from, the rights conveyed by a patent, the FDA can provide market exclusivity for an approved drug. Two years of exclusivity are extended to drugs in clinical testing when the 1984 Act was passed. The FDA also will not consider applications for a generic version of a new chemical entity for five years after approval of the original. This applies even if there is no patent on the drug. According to CBO, however, this may, in actuality, add more than five years because abbreviated drug applications often take more than 30 months, on average, for approval. Added together, this may provide over seven years of market exclusivity. The Food and Drug Administration also is permitted to grant a three year exclusivity period if a new drug application (or supplemental application) necessitates additional clinical investigation. These situations include new dosage forms for already approved drugs, a new use for a drug, or for over-the-counter marketing of a drug. This market exclusivity only pertains to the new indication and does not prevent the approval of a new pharmaceutical if all the required clinical studies are performed to support the same changes.[194] The intent is to encourage ongoing innovation on existing pharmaceuticals.

Another mechanism established by the 1984 Act extends market exclusivity if the FDA accepts a new claim for an existing pharmaceutical. For example, Bristol-Myers Squibb repositioned Excedrin as Excedrin Migraine with the same active ingredients. Similarly, JandJ/McNeil produces Motrin Migraine Pain as well as Motrin.[195] The argument has been made that the brand name drug companies are creating "improved drug entities" based on their original invention. When approved by the FDA, the changes made permit three years of exclusivity on the marketing of the pharmaceutical if a new patent is not forthcoming and an additional 20 years if a patent issues. If the original drug is removed from the market, however, a generic for that pharmaceutical cannot be introduced.[196] Allowing this removal to occur, CBO argues, can prevent generics from coming to market.

Assessing the effect of such provisions, the Congressional Budget Office's 1998 study indicated that the 1984 Act provided brand name drugs with an additional 2.8 years of market exclusivity prior to the entry of generics (including drugs that did not obtain an extension under the terms of the 1984 Act). However, Grabowski and Vernon found that the extent of overall market exclusivity for new drugs has actually decreased in contrast to the situation prior to implementation of the 1984 Act.[197] For example, according to PhRMA, while Inderal, introduced in 1965, experienced 10 years of market exclusivity and Tagamet, introduced in 1977, had six years of market exclusivity, Diflucan, introduced in 1990, received only two years of exclusivity and Invirase, introduced in 1995, had just three months on the market before a generic was introduced.[198]

Despite the ability of the FDA to offer market exclusivity, some experts argue that the 1984 Act ". . .has also significantly curtailed the expected revenues to innovative firms from the latter phases of their drug's life cycle."[199] According to CBO, despite this period of exclusivity, most of the average cost of drug development cannot be recouped. CBO found that the increase in generics has led to an average $27 million (or 12%) decrease in the total return to a new drug (not including antibiotics not covered by the 1984 Act). The "average market price" declines even though the cost of the innovator drug increases because generics make up a larger share of the market.[200] This has occurred at the same time that R&D costs and time to market have increased.[201]

In order to compete with other companies, brand name firms may bring out generic versions of their own drugs before the original patent expires. The intent is to be the first to market and to establish market advantage with pharmacies which "...usually buy the first low-cost alternative, then rarely switch to other brands once customers get used to it." This occurs despite some evidence that the brand name firms price their generics at 10 to25% less than the original drug in contrast to other generic products that typically cost half as much.[202] Upjohn, upon introducing a generic version of Xanax one month before the patent expired, soon controlled 90% of the generic market for similar drugs.[203] However, Syntex, which brought out a generic version of its drug Naprosyn two months prior to patent expiration and initially captured three-quarters of the generic market, found it lost almost two-thirds of this market when other generics were introduced.[204]

Research by Kamien and Zang published in 1999 states that brand name company introduction of generic substitutes ". . .appears to benefit both them and the consumers." Profits increase for these firms above and beyond that which could be made solely with the original drug. This action also allows the firm to raise prices on the innovator pharmaceutical. According to Kamien and Zang, consumers are better off because brand name generics provide a lower cost alternative before the original patent expires, even though this benefit only lasts for a month or two. However, once the patent expires, the brand name company obtains a "first-mover" advantage on the marketplace. At this point, the average price of the brand name and generic drug is lower because of competition. Thus, these two authors argue, the producers of generic drugs are worse off in this situation than both the brand name firms and the public.[205]

Possible Issues and Potential Concerns[206]

Given the increasing investment in research and development and the gains in research intensity of the pharmaceutical industry, it appears that the 1984 Act has not deterred the search foR&Development of new drugs. In assessing the effects of the 1984 Act, the Congressional Budget Office found that "[o]verall, it appears that the incentives for drug companies to innovate have remained intact since. . ." the passage of the legislation. While brand name companies have experienced some loss due to the increased competition following patent expiration, the extension of patent terms that has resulted from the implementation of the 1984 Act has matched the ". . . average three-year delay between patent expiration and generic entry that existed before the act (in cases where generic entry occurred)." The report concludes:

> Still, those extensions played an important role in protecting the returns from drug companies' research and development. Without them, the rise in generic market share since 1984 would have dramatically lowered the expected returns from marketing a drug and might have caused the pharmaceutical industry to reduce its investment in R&D. In that case, a successful innovator drug would have been likely to lose over 40 percent of its market to generic competitors just after reaching its peak year in sales. If the pre-1984 level of R&D investment was desirable, then the patent extensions benefitted society by preserving most of the returns from marketing a new drug.

On the other hand, some experts argue that the large and growing private and public investment in pharmaceutical research and development makes it ". . . clear that the patent-related provisions of the '84 Act are no longer necessary to achieve the policy of fostering innovation while insuring public access to older drugs at competitive prices."[207] According to this view, such provisions permit and encourage manipulation. Elimination of patent extension and market exclusivity, such critics maintain, would allow the market to operate at "maximum efficiency."[208] The Congressional Budget Office points out that accelerating FDA review process (by one year) would be more helpful to innovator drugs than providing patent extension. Their research indicates that "the patent extensions available under the [1984 Act] were not sufficient to fully preserve the returns from marketing new brand-name drugs." Shortening the process in FDA by one year, however would provide a net benefit of approximately $22 million for one drug.[209]

Congressional interest in the 1984 Act continues.[210] In further exploring the topic, the Congress is likely to consider various issues surrounding implementation of the legislation. Highlighted below are possible areas for discussion. Among the concerns is whether or not the environment in which the original law was enacted still exists and if adjustments should be made to reflect any changes. Has the implementation of the 1984 Act led to any new, unanticipated benefits or consequences? Of fundamental interest is whether or not the goals and incentives contained in the law remain valid after 18 years.

In assessing the current environment within which the provisions of the 1984 Act are applied, an important question is whether or not the state of the FDA approval process remains the same as when the original legislation was passed. At the time Congress originally debated the law, the average FDA drug approval time was over 30 months. In 1999, this period had dropped by more than half. The Food and Drug Administration maintains that the mean approval time in 1999 was 12.6 months.[211] Concurrently, the number of clinical studies required per new drug application has increased. At issue is whether or not the patent term extension provisions and market exclusivity provisions contained in the 1984 Act accurately reflect the delays associated with the FDA approval process as it operates today.

In the first session, the 106th Congress enacted the American Inventors Protection Act (P.L. 106-113). This legislation requires that certain deadlines be met by the Patent and Trademark Office in the issuance of a patent. Among these deadlines are 14 months for the first office action, four months for a subsequent action, and four months between payment of an issuance fee and the grant of a patent. The original patent application must be completed within three years of actual filing except if the delays resulted from continuing applications and appeals on behalf of the filing party. If these time constraints are not adhered to, the patent holder may receive a day-for-day extension of the patent term. How might this new law affect the implementation and impact of the 1984 Act?

Since the passage of the 1984 Act, Congress has created additional market exclusivity provisions for certain drugs. The Orphan Drug Act provides a company the exclusive right to market a drug that has been properly designated (to address diseases that affect less than 2,000 people annually) for seven years from the date of FDA approval. In addition, the 1997 FDA Modernization Act, as amended by the Best Pharmaceuticals for Children Act, extends market exclusivity for six months if companies undertake studies on the use of a drug in children. Do these laws affect the balance between encouraging innovation and encouraging the introduction of generics promoted by the 1984 Act?

The environment within which pharmaceutical research and development are performed has changed. The costs of R&D have increased; according to DiMasi, R&D costs have shown a 10% compounded annual growth rate.[212] This is reflected in an increase in the R&D intensity of the industry. In 1980, R&D expenditures were 11.9% of sales by research-based pharmaceutical companies; for 2001, it is estimated that R&D will increase to 17.7% of sales (although down from 20.3% in 2000).[213] The use of collaborative partnerships has expanded to help reduce costs. Similarly, there have been an increased number of mergers among pharmaceutical companies. These activities have occurred as the importance of "blockbuster" drugs to a company has increased. Today, the blockbuster drugs a firm develops are its principle source of profits; 10% of drugs account for approximately 80% of global sales.[214] It is the expiration of patents on these blockbuster drugs that typically draw the most attention. Given the current R&D environment within which pharmaceutical companies operate, do the provisions of the 1984 Act provide the necessary incentives for further innovation?

The biotechnology industry was in its infancy during the period that the 1984 Act was debated and passed. Therefore, some experts argue, the provisions of the law are not relevant to biotechnology products that are an increasing component of the drug industry. The ownership of intellectual property is particularly important to biotechnology companies. The

U.S. biotechnology industry is one of the most research-intensive sectors in the world as it committed $9.9 billion to R&D in 1998. However, these firms are typically small and do not yet have profits to finance additional R&D. According to the Biotechnology Industry Organization, most of these companies finance research and development from equity capital not profits. Only 3.5% of biotech firms have sales; therefore most depend on venture capital and IPOs to support on-going R&D.[215] Industry sources maintain that patents are a necessity for raising this equity capital.[216] Biotechnology products involve the growth of a biological component, rather than the development of a chemically synthesized component[217] and some observers believe that the abbreviated bioequivalent determination established under the 1984 Act is not appropriate.[218] Biotech drugs may be similar in their chemical or biological make-up but test differently in clinical trials.[219] Based on these factors, some in the industry maintain there is a need to create regulations similar to those in the original Act for biologics in order to develop a generic sector such that exists in pharmaceuticals.[220]

The 1984 Act provides rewards for certain activities as discussed above. This leads to concerns over whether or not such a system, while encouraging certain positive efforts, also leads to less beneficial company policies and practices. How do patent term extensions and market exclusivity provisions encourage and/or facilitate activities by firms that might not foster innovation? For example, the law provides the opportunity to extend market exclusivity by listing patents in the Orange Book.[221] Some experts argue that this has encouraged firms to list patents for products that are not considered marketable.[222] Others maintain that companies increase the number of patents associated with a particular drug to prevent the introduction of generics. The structure of the patent portfolio for a new drug may reflect the provisions of the 1984 Act; how are the traditional process of research, development, and commercialization affected by considerations of future claims under the law?

Other concerns have been expressed regarding allegations that brand name firms are paying companies not to bring generics to market. Originally, the FDA required that a generic company that filed an abbreviated new drug application (ANDA) had to be sued for patent infringement and win in court before the agency would offer the 180 days of market exclusivity. FDA guidelines developed in 1998, eliminated the necessity for a "successful defense" by a generic manufacturer against claims of patent infringement prior to receiving the 180 day market exclusivity. The intent of this provision had been to provide an incentive for marketing a generic to recover litigation costs and make full use of the exclusivity provided.[223] However, now the only criteria for market exclusivity is receiving the first to file position. This has led, it is argued, to the filing of ". . . substandard or 'sham' ANDAs as generic companies race to establish themselves as being the first to file."[224] As the regulations now stand, the 180 days is triggered by the commercial marketing of the generic. Some experts maintain that this change allows for activities that conflict with the intent of the law. It has been alleged that certain brand name manufacturers have paid the generic firms granted exclusivity not to begin selling their products so as not to open the market to other generics. The Federal Trade Commission brought suit against Hoechst Marion Roussel and Andrx Pharmaceuticals and a federal judge declared that the firms violated antitrust laws when Hoechst paid Andrx to delay marketing of their generic version of the brand name drug. Another similar case involves Abbott Laboratories and Geneva Pharmaceuticals. The FTC takes a case-by-case approach to the antitrust issue. The companies involved in these situations argue that the agreements are a result of patent disputes, not a means to block market access.[225]

Addressing the above concerns may provide the context within which to assess the means by which the 1984 Act has attempted to achieve congressional intent. Are patent extensions and market exclusivity provisions the most effective and/or efficient means to encourage innovation or do other mechanisms exist? Are the existing incentives for the development and marketing of generic drugs the most productive way to offer lower-cost pharmaceuticals to the public? Are they necessary in today's environment? Does the argument that patent expiration, not patent extension, stimulates innovation figure into the discussion? How does the finding by CBO regarding the savings to be achieved by reducing FDA approval time affect an assessment of the results of the implementation of the 1984 Act?

A more fundamental issue that might be explored is whether or not the goals and incentives in the law remain valid within the present environment? Have the legal reforms served to encourage the introduction of lower-cost generic drugs while simultaneously providing incentives to further pharmaceutical innovation? In light of current events, is the effort to balance these objectives still appropriate and/or necessary?

REFERENCES

[1] For additional discussion of the Hatch-Waxman Act, including recent legislative activity see CRS Report RL32377, *The Hatch-Waxman Act: Legislative Changes Affecting Pharmaceutical Patents*, by Wendy H. Schacht and John R. Thomas; and CRS Report RL31379, *The "Hatch-Waxman" Act: Selected Patent-Related Issues*, by Wendy H. Schacht and John R. Thomas.

[2] Bale Jr., Harvey E., "Patent Protection and Pharmaceutical Innovation," 29 *New York University Journal of International Law and Politics* (1996-97), 95.

[3] Bale, Jr., Harvey E., "The conflicts between parallel trade and product access and innovation: the case for pharmaceuticals," 1 *Journal of International Economic Law* (1998), 637, 641.

[4] John B. Shoven, "Intellectual Property Rights and Economic Growth," in eds. Charls Walker and Mark A. Bloomfield, *Intellectual Property Rights and Capital Formation in the Next Decade*, (New York, University Press of America, 1988), 46.

[5] Robert P. Benko, "Intellectual Property Rights and New Technologies," in Walker, et. al., Intellectual Property Rights and Capital Formation in the Next Decade, 27.

[6] Stanley M. Besen and Leo J. Raskind, "An Introduction to the Law and Economics of Intellectual Property," *Journal of Economic Perspectives*, Winter 1991, 5.

[7] For a list of relevant research in this area see Council of Economic Advisors. *Supporting Research and Development to Promote Economic Growth: The Federal Government's Role*, (October 1995), 6-7.

[8] David J. Teece, "Profiting from Technological Innovation: Implications for Integration, Collaboration, Licensing, and Public Policy," in *The Competitive Challenge,* ed. David J. Teece, (Cambridge: Ballinger Publishing Co., 1987), 188.

[9] Edwin Mansfield. "Intellectual Property Rights, Technological Change, and Economic Growth," in *Intellectual Property Rights and Capital Formation in the Next Decade*,

eds. Charls E. Walker and Mark A. Bloomfield (New York: University Press of America, 1988), 10-11.

[10] Kenneth W. Dam, "The Economic Underpinnings of Patent Law," *Journal of Legal Studies*, January 1994, 247.

[11] Robert P. Merges, "Commercial Success and Patent Standards: Economic Perspectives on Innovation," *California Law Review*, July 1988, 876.

[12] Dam , The Economic Underpinnings of Patent Law, 266-267. Scope is determined by the number of claims made in a patent. Claims are the technical descriptions associated with the invention. In order for an idea to receive a patent, the law requires that it be ". . .new, useful [novel], and nonobvious to a person of ordinary skill in the art to which the invention pertains." See footnote 12, p. 7.

[13] Robert P. Merges and Richard R. Nelson, "On the Complex Economics of Patent Scope," Columbia Law Review, May 1990, 908.

[14] Richard C. Levin and Alvin K. Klevorick, Richard R. Nelson, Sidney G. Winter. "Appropriating the Returns for Industrial Research and Development," *Brookings Papers on Economic Activity*, 1987, in *The Economics of Technical Change*, eds. Edwin Mansfield and Elizabeth Mansfield (Vermont, Edward Elgar Publishing Co., 1993), 254.

[15] Ibid., 255 and 257. See also: Mansfield, Intellectual Property Rights, Technological Change, and Economic Growth,12 and 13.

[16] Levin, et. al., Appropriating the Returns for Industrial Research and Development, 253.

[17] Ibid., 269.

[18] Edwin Mansfield, Mark Schwartz, and Samuel Wagner. "Imitation Costs and Patents: An Empirical Study," *The Economic Journal*, December 1981, in *The Economics of Technical Change*, 270.

[19] 35 U.S.C. § 101.

[20] See Diamond v. Chakrabarty, 447 U.S. 303 (1980).

[21] See In re Pleuddemann, 910 F.2d 823 (Fed. Cir. 1990).

[22] 35 U.S.C. § 100(b).

[23] See Titanium Metals Corp. v. Banner, 778 F.2d 775 (Fed. Cir. 1985).

[24] 35 U.S.C. § 101.

[25] *See Mitchell v. Tilghman*, 86 U.S. (19 Wall.) 287, 396 (1873).

[26] 385 U.S. 419 (1966).

[27] Ibid., p. 534-35.

[28] Ibid., p. 535.

[29] Ibid., p. 536 (quoting *Application of Ruschig*, 343 F.2d 965, 970 (CCPA 1965)).

[30] 51 F.3d 1560 (Fed. Cir. 1995).

[31] Ibid., p. 1565.

[32] Ibid ., p. 1568.

[33] Machin, Nathan. Prospective Utility: A New Interpretation of the Utility Requirement of Section 101 of the Patent Act, 87 *California Law Review* (1999), p. 421, 432.

[34] 51 F.3d at 1564-68.

[35] 35 U.S.C. § 102.

[36] 35 U.S.C. § 103(a).

[37] 37 C.F.R. § 1.16.

[38] 35 U.S.C. § 112. 1.

[39] See Atlas Powder Co. v. E.I. DuPont de Nemours and Co., 750 F.2d 1569 (Fed. Cir. 1984).
[40] *See Vas-Cath Inc. v. Marhurkar*, 935 F.2d 1555 (Fed. Cir. 1991).
[41] *See Glaxo Inc. v. Novopharm Ltd.*, 52 F.3d 1043 (Fed. Cir. 1995).
[42] 35 U.S.C. § 112. 2.
[43] See Orthokinetics, Inc. v. Safety Travel Chairs, Inc., 806 F.2d 1565 (Fed. Cir. 1986).
[44] 37 C.F.R. § 1.56.
[45] Levine, Russell E., et al., "Ex Parte Patent Practice and the Rights of Third Parties," 45 American University Law Review (1996), 1987.
[46] 35 U.S.C. § 132.
[47] *See In re Oetiker*, 977 F.2d 1443 (Fed. Cir. 1992).
[48] 35 U.S.C. §§ 120, 133, 134.
[49] American Inventors Protection Act of 1999, Pub. L. No. 106-113.
[50] 35 U.S.C. § 122.
[51] 35 U.S.C. § 135.
[52] 35 U.S.C. § 102(g).
[53] 35 U.S.C. § 251.
[54] 35 U.S.C. § 302.
[55] 35 U.S.C. § 303.
[56] 35 U.S.C. § 304.
[57] 35 U.S.C. § 307.
[58] American Inventors Protection Act of 1999, Pub. L. No. 106-113. *See* Janis, Mark D., "Inter Partes Reexamination," 10 Fordham Intellectual Property, Media and Entertainment Law Journal (2000), 481.
[59] 35 U.S.C. §§ 252, 307(a).
[60] 35 U.S.C. § 154(a).
[61] *See* Lemley, Mark A. An Empirical Study of the Twenty-Year Patent Term, 22 *American Intellectual Property Law Association Quarterly Journal* (1994), p. 369.
[62] Pub. L. No. 98-417, 98 Stat. 1585 (1984).
[63] 37 C.F.R. § 1.16.
[64] 35 U.S.C. § 271(a).
[65] *Johnston v. IVAC Corp.*, 885 F.2d 1574, 1580 (Fed.Cir.1989).
[66] 733 F.2d 858 (Fed. Cir. 1984).
[67] 35 U.S.C. § 271(b), (c).
[68] See Chimuinatta Concrete Concepts, Inc. v. Cardinal Industries, Inc., 145 F.3d 1303 (Fed. Cir. 1998).
[69] Graver Tank v. Linde Air Products Co., 339 U.S. 605 (1950).
[70] 520 U.S. 17 (1997).
[71] *See Jurgens v. CBK, Ltd.*, 80 F.3d 1566, 1572 n.2 (Fed. Cir. 1996).
[72] Leatherman Tool Group Inc. v. Cooper Industries Inc., 131 F.3d 1011, 1015 (Fed. Cir. 1996).
[73] See Bio-Technology General Corp. v. Genentech Inc., 80 F.3d 1553, 1559 (Fed. Cir.), cert. denied, 117 S. Ct. 274 (1996).
[74] See Dowagiac Mfg. Co. v. Minnesota Moline Plow Co., 35 U.S. 641, 650 (1915).
[75] *See Intel Corp. v. ULSI Corp.*, 995 F.2d 1566 (Fed. Cir. 1993), *cert. denied*, 510 U.S. 1092 (1994).

[76] See Joy Technologies, Inc. v. Flakt, Inc., 6 F.3d 770 (Fed. Cir. 1993).
[77] Pub. L. No. 100-418.
[78] 35 U.S.C. § 271(g).
[79] 35 U.S.C. § 271(g)(1), (2). See Eli Lilly and Co. v. American Cyanamid Co., 82 F.3d 1568 (Fed. Cir. 1996).
[80] 35 U.S.C. § 287(b).
[81] *Rohm and Haas Co. v. Brotech Corp.*, 127 F.3d 1089 (Fed. Cir. 1997).
[82] 35 U.S.C. § 295.
[83] 35 U.S.C. § 281.
[84] 35 U.S.C. § 282.
[85] The Federal Courts Improvement Act of 1982, Pub. L. No. 97-164, 96 Stat. 25 (1982).
[86] 35 U.S.C. § 283.
[87] See Mentor Graphics Corp. v. Quickturn Design Systems, Inc., 150 F.3d 1374, 1377, 47 USPQ2d 1683, 1685 (Fed. Cir. 1998).
[88] 35 U.S.C. § 284.
[89] 35 U.S.C. § 286.
[90] See General Motors Corp. v. Devex Corp., 461 U.S. 648 (1983).
[91] 35 U.S.C. § 261.
[92] Ibid.
[93] *Spindelfabrik Suessen-Schurr v. Schubert and Salzer*, 829 F.2d 1075, 1081 (Fed.Cir. 1987), *cert. denied*, 484 U.S. 1063 (1988).
[94] Adelman, Martin J., et al., *Cases and Materials on Patent Law* (1998), 1231.
[95] Pub. L. No. 98-417, 98 Stat. 1585 (1984).
[96] Rea, Teresa Stanek, "Striking the Right Balance Between Innovation and Drug Price Competition: Understanding the Hatch-Waxman Act–An Introduction of Speakers," 54 Food Drug Law Journal (1999), 223, 224.
[97] *See, e.g., Glaxo, Inc. v. Novopharm, Ltd.*, 110 F.3d 1562, 1568 (Fed. Cir. 1997).
[98] *See, e.g.,* McGough, Kevin J., "Preserving the Compromise: The Plain Meaning of Waxman-Hatch Exclusivity," 45 *Food, Drug and Cosmetic Law Journal* (1990), 487.
[99] 35 U.S.C. § 271(a).
[100] Chisum, Donald S., *Principles of Patent Law* (Foundation Press, New York, New York, 1998), 5.
[101] 21 U.S.C. § 355(b). Prior to 1962, the drug approval process was solely directed towards safety. See Mossinghoff, Gerald J., "Overview of the Hatch-Waxman Act and Its Impact on the Drug Development Process," 54 *Food and Drug Law Journal* (1998), 187.
[102] See *In re Brana*, 51 F.3d 1560 (Fed. Cir. 1995).
[103] 21 U.S.C. § 355(b). Prior to 1962, the drug approval process was solely directed towards safety. See Mossinghoff, *supra* note 8, at 187.
[104] Engelberg, Alfred B., "Special Patent Provisions for Pharmaceuticals: Have They Outlived Their Usefulness?," 39 *IDEA: Journal of Law and Technology* (1999), 389, 396. Generic drugs are versions of brand-name prescription drugs that are often sold without a trademark and that contain the same active ingredients, but not necessarily the same inactive ingredients, as the original. *United States v. Generix Drug Co.*, 460 U.S. 435, 455 (1983).

[105] Buchanan, J. Matthew, "Medical Device Patent Rights in the Age of FDA Modernization: the Potential Effect of Regulatory Streamlining on the Right to Exclude," 30 *University of Toledo Law Review* (1999) 305, 316.
[106] Engelberg, supra note 103, at 396-97.
[107] Buchanan, *supra* note 104.
[108] 35 U.S.C. § 271(a).
[109] 29 F.Cas. 1120, 1121 (C.C.Mass. 1813)(No. 17,600).
[110] 19 F.Cas. 1048, 1049 (C.C.S.D.N.Y. 1861) (No. 11,279).
[111] *See* Note, "Experimental Use as Patent Infringement: The Impropriety of a Broad Exception," 100 *Yale Law Journal* (1991), 2169.
[112] Bee, Richard E., "Experimental Use as An Act of Patent Infringement," 39 *Journal of the Patent Office Society* (1957), 357.
[113] 733 F.2d 858 (Fed. Cir. 1984).
[114] *See* U.S. Patent No. 3,299,053 ("Novel 1 and/or 4-substituted alkyl 5-aromatic-3H-1,4-benzodiazepines and benzodiazepine-2-ones.").
[115] 572 F. Supp. 255 (E.D.N.Y. 1983).
[116] 35 U.S.C. § 271(a).
[117] 733 F.2d at 862-64.
[118] 733 F.2d at 863.
[119] Pub. L. No. 75-717, 52 Stat. 1040 (1938) (codified as amended 21 U.S.C. §§ 301 et seq.).
[120] Pub. L. No. 82-593, 66 Stat. 792 (1952) (codified as amended 35 U.S.C. § 1 et seq.).
[121] 733 F.2d at 863-64.
[122] 733 F.2d at 864-66.
[123] 733 F.2d at 865-67.
[124] The specific provisions are 15 U.S.C. §§ 68b-68c, 70b; 21 U.S.C. §§ 301, 355, 360cc; 28 U.S.C. § 2201; and 35 U.S.C. §§ 156, 271, 282.
[125] Glover, Gregory J., "Regulatory Concerns and Market Exclusivity," *Health Care MandA 2000*, 1175 Practising Law Institute (2000), 629, 633.
[126] 21 U.S.C. § 355(j).
[127] Ibid.
[128] 35 U.S.C. § 156. Prior to United States adherence to the World Trade Organization, patents were granted a term of 17 years from the date of issuance. On June 8, 1995, the effective patent term was changed to 20 years measured from the date the patent application was filed. Patents in existence as of June 8, 1995, or patents that issued from applications pending at the USPTO as of the date, have a term equal to the greater of 17 years from issuance or 20 years from grant.
[129] 35 U.S.C. § 156.
[130] 35 U.S.C. § 156(c).
[131] 35 U.S.C. § 156(d)(2)(B).
[132] Tyson, Karin L., "The Role of the Patent and Trademark Office Under 35 U.S.C. Section 156," 54 Food and Drug Law Journal (1999), 205.
[133] 21 U.S.C. § 355(j)(4)(D)(I).
[134] 21 U.S.C. § 355(j)(4)(D)(iii).
[135] 21 U.S.C. § 355(j)(D)(ii).
[136] Glover, supra note 124, at 634.

[137] See supra notes 112-122 and accompanying text.
[138] 775 F. Supp. 1269 (N.D. Cal.), *affirmed*, 991 F.2d 808 (Fed. Cir. 1993).
[139] Ibid at 1283.
[140] Ibid at 1284.
[141] 496 U.S. 661 (1990).
[142] *See* Engelberg, *supra* note 103, at 402.
[143] 21 U.S.C. § 355(b)(1), 355(j)(2)(A)(vi).
[144] Food and Drug Administration, Center for Drug Evaluation and Research, Approved Drug Products with Therapeutic Equivalence Evaluations; Dickinson, Elizabeth A., "FDA's Role in Making Exclusivity Determinations," 54 *Food and Drug Law Journal* (1999), 195, 196.
[145] Mossinghoff, supra note 100, at 189.
[146] 21 U.S.C. §§ 355(j)(5)(A), (B)(I).
[147] 21 U.S.C. § 355(c)(3)(C).
[148] 35 U.S.C. §§ 271(e)(4)(A).
[149] 21 U.S.C. §§ 355(j)(5)(B)(iii)(I)(III).
[150] Dickinson, *supra* note 143, at 199.
[151] 140 F.3d 1060 (D.C. Cir. 1998).
[152] 140 F.3d at 1069.
[153] 21 U.S.C. § 355(j)(5)(B)(iv)(I), (II).
[154] Pub. L. No. 100-670, 102 Stat. 3971 (1988).
[155] Pub. L. No. 103-465, 108 Stat. 4809 (1994).
[156] *See Bristol-Myers Squibb v. Royce*, 69 F.3d 1130 (Fed. Cir. 1995).
[157] Mossinghoff, *supra* note 100, at 188.
[158] Pub. L. No. 97-414, 96 Stat. 2049 (1983) (codified at 21 U.S.C. § 360aa et seq.).
[159] Pub. L. No. 105-115, 111 Stat. 2296 (1997) (codified at 28 U.S.C. § 352(a)).
[160] Dickinson, *supra* note 143, at 201-03.
[161] Ibid.
[162] Karst, Kurt R., "Pediatric Testing of Prescription Drugs: The Food and Drug Administration's Carrot and Stick for the Pharmaceutical Industry," 49 *American University Law Review* (2000), 739, 750.
[163] Ibid at 203.
[164] Glover, *supra* note 124.
[165] U.S. Department of Commerce and McGraw-Hill. U.S. Trade and Industry Outlook 2000 (Washington, D.C., McGraw-Hill, 1999), 11-16.
[166] Ibid., 11-14.
[167] National Science Foundation. *Science and Engineering Indicators 2002*, available at [http://www.nsf.gov].
[168] Pharmaceutical Research and Manufacturers of America. Pharmaceutical Industry *Profile 2004,* available at [http://www.phrma.org].
[169] National Institutes of Health website, available at [http://grants.nih.gov/grants/award].
[170] Iain Cockburn, Rebecca Henderson, Luigi Orsenigo, and Gary P. Pisano. "Pharmaceuticals and Biotechnology," in ed. David C. Mowery, *U.S. Industry in 2000* (National Academy Press, Washington, D.C. 1999), 363.
[171] Michael P. Ryan. Knowledge Diplomacy, Global Competition and the Politics of Intellectual Property (Washington, D.C., Brookings Institution Press, 1998), 30.

[172] Pharmaceutical Industry Profile 2004.
[173] Henry G. Grabowski and John M. Vernon. "Brand Loyalty, Entry, and Price Competition in Pharmaceuticals After the 1984 Drug Act," *Journal of Law and Economics*, October 1992, 334.
[174] Congressional Budget Office. How Increased Competition from Generic Drugs has Affected Prices and Returns in the Pharmaceutical Industry (Washington, D.C., July 1998) available at [http://www.cbo.gov].
[175] F.M. Scherer. *Industry Structure, Strategy, and Public Policy* (New York, HarperCollins, 1996), 373.
[176] Ibid., 376.
[177] Information posted on [http://www.phrma.org].
[178] Information available at [http://www.gphaonline.org].
[179] Ibid.
[180] Brand Loyalty, Entry, and Price Competition in Pharmaceuticals After the 1984 Act, 347.
[181] Ibid., 347.
[182] Alfred B. Engelberg. "Special Patent Provisions for Pharmaceuticals: Have They Outlived Their Usefulness?," *IDEA: The Journal of Law and Technology*, 1999.
[183] Henry G. Grabowski. *The Effect of the 1984 Hatch-Waxman Act on Generic Competition and Drug Innovation*. Testimony before the Senate Committee on the Judiciary, March 5, 1996.
[184] Brand Loyalty, Entry, and Price Competition in Pharmaceuticals After the 1984 Drug Act, 347.
[185] Morton I. Kamien and Israel Zang. "Virtual Patent Extension by Cannibalization," Southern Economic Journal, July 1999.
[186] Industry Structure, Strategy, and Public Policy, 377.
[187] Ibid., 379 summarizing the work reported in Richard E. Caves, Michael D. Whinston, Mark A. Hurwitz. "Patent Expiration, Entry, and Competition in the U.S. Pharmaceutical Industry," *Brookings Papers*, 1991.
[188] Brand Loyalty, Entry, and Price Competition in Pharmaceuticals After the 1984 Drug Act, 341.
[189] Ibid., 340.
[190] For a detailed description of these provisions see Introduction to the Drug Price Competition and Patent Term Restoration Act of 1984.
[191] The Effects of the 1984 Hatch-Waxman Act on Generic Competition and Drug Innovation.
[192] Marilyn Werber Serafini. "The Price of Miracles," *National Journal*, March 25, 2000.
[193] Shailagh Murray. "Senate Mulls Bill to Extend Drug Patents," *The Wall Street Journal*, August 5, 1999.
[194] Elizabeth H. Dickinson. "FDA's Role in Making Exclusivity Determinations," *Food and Drug Law Journal*, 1999, 201.
[195] Christine Bittar. "As Patents Expire, Look for Extensions," *Brandweek*, June 19, 2000, 98.
[196] Feliza Mirasol. "Generic Drug Industry Faces Regulatory and Patent Issues," *Chemical Market Reporter*, April 12, 1999.

[197] The Effects of the 1984 Hatch-Waxman Act on Generic Competition and Drug Innovation.
[198] Information available at [http://www.phrma.org]
[199] Brand Loyalty, Entry, and Price Competition in Pharmaceuticals After the 1984 Drug Act, 347.
[200] Ibid., 335.
[201] The Effects of the 1984 Hatch-Waxman Act on Generic Competition and Drug Innovation.
[202] Catherine Yang. "The Drugmakers vs. the Trustbusters," *Business Week*, Setermber 5, 1994.
[203] Virtual Patent Extension by Cannibalization.
[204] The Drugmakers vs. the Trustbusters.
[205] Virtual Patent Extension by Cannibalization.
[206] Note that changes in the law were subsequently made by P.L. 108-173, the Medicare Prescription Drug and Modernization Act of 2003. For more information on this legislation see CRS Report RL32377, *The Hatch-Waxman Act: Legislative Changes Affecting Pharmaceutical Patents*, by Wendy H. Schacht and John R. Thomas.
[207] Special Patent Provisions for Pharmaceuticals: Have They Outlived Their Usefulness?
[208] Ibid.
[209] How Increased Competition from Generic Drugs has affected Prices and Returns in the Pharmaceutical Industry.
[210] For additional discussion see CRS Report RL31379 and CRS Report RL32377.
[211] Under the provisions of the 1992 Prescription Drug User Fee Act pharmaceutical companies are charged to have certain new drug applications approved by the FDA. It is expected that the FDA will complete the approval process within a specified time frame.
[212] Joseph A. DiMasi. Presentation at a meeting on *Innovation in the Pharmaceutical Industry: New Evidence on Structure, Process, and Outcomes* held by the American Enterprise Institute, October 6, 2000.
[213] [http://www.phrma.org]
[214] Henry G. Grabowski. Presentation at a meeting on *Innovation in the Pharmaceutical Industry: New Evidence on Structure, Process, and Outcomes* held by the American Enterprise Institute, October 6, 2000.
[215] Biotechnology Industry Organization, "Contributions to New Medicine from Government-Funded Basic Biomedical Research," testimony submitted to the Senate Appropriations Subcommittee on Labor, Health, and Human Services, Education, and Related Agencies, April 1999, [http://www.bio.org].
[216] Adriel Bettleheim, "Drugmakers Under Siege," *CQ Outlook*, September 25, 1999, 10.
[217] The Price of Miracles.
[218] David Schmickel. "The Biotechnology Industry Organization's View on Hatch-Waxman Reform, *Food and Drug Law Journal*, 1999, 242.
[219] Ibid., 241-242.
[220] The Price of Miracles.
[221] For a discussion of the Orange Book see Introduction to the Drug Price Competition and Patent Term Restoration Act of 1984.
[222] Special Patent Provisions for Pharmaceuticals: Have They Outlived Their Usefulness?

[223] "New FDA Guidance Raises Anticompetutuve Concerns, Says National Association of Pharmaceutical Manufacturers," *Business Wire*, June 23, 1998.
[224] Generic Drug Industry Faces Regulatory and Patent Issues.
[225] "Driving Up Drug Prices," *New York Times*, July 26, 2000.

Chapter 3

INDUSTRIAL COMPETITIVENESS AND TECHNOLOGICAL ADVANCEMENT: DEBATE OVER GOVERNMENT POLICY[*]

Wendy H. Schacht

ABSTRACT

There is on-going interest in the pace of U.S. technological advancement due to its influence on U.S. economic growth, productivity, and international competitiveness. Because technology can contribute to economic growth and productivity increases, congressional interest has focused on how to augment private-sector technological development. Legislative activity over the past two decades has created a policy for technology development, albeit an ad hoc one. Because of the lack of consensus on the scope and direction of a national policy, Congress has taken an incremental approach aimed at creating new mechanisms to facilitate technological advancement in particular areas and making changes and improvements as necessary.

Congressional action has mandated specific technology development programs and obligations in federal agencies that did not initially support such efforts. Many programs were created based upon what individual committees judged appropriate within the agencies over which they had authorization or appropriation responsibilities. The use of line item funding for these activities, including the Advanced Technology Program and the Manufacturing Extension Program of the National Institute of Standards and Technology, as well as for the Undersecretary for Technology at the Department of Commerce, is viewed as a way to ensure that the government encourages technological advance in the private sector.

Some legislative activity, beginning in the 104th Congress, has been directed at eliminating or significantly curtailing many of these federal efforts. Although this approach has not been successful, the budgets for several programs have declined. Questions have been raised concerning the proper role of the federal government in technology development and the competitiveness of U.S. industry. As the 109th Congress

---
[*] Excerpted from CRS Report IB91132, dated April 1, 2005.

develops its budget priorities, how the government encourages technological progress in the private sector again may be explored and/or redefined.

MOST RECENT DEVELOPMENTS

Over the past 25 years, congressional initiatives have supported technological advancement in U.S. industry. This approach has involved both direct measures that concern budget outlays and the provision of services by government agencies (such as the Advanced Technology Program (ATP) and the Manufacturing Extension Partnership (MEP) of the National Institute of Standards and Technology) and indirect measures that include financial incentives and legal changes. Many of these efforts, however, have been revisited since the 104th Congress given the Republican majority's statements in favor of indirect strategies such as tax policies, intellectual property rights, and antitrust laws to promote technological advancement; increased government support for basic research; and decreased direct federal funding for private sector technology initiatives. While no program has been eliminated, several have been financed at reduced levels. Beginning in FY2000, the original House-passed appropriation bills did not include funding for ATP. In addition, the President's FY2003 budget first requested a significant reduction in support for MEP based on the idea that all manufacturing extension centers operating more than six years should continue without federal funding. In the previous Congress, P.L. 108-199, the FY2004 Consolidated Appropriations Act, reduced support for MEP 63%, providing $38.7 million for the program. The legislation also financed ATP at $170.5 million (including the mandated 0.59% rescission). The FY2005 Omnibus Appropriations Act, P.L. 108-447, provided ATP with $136.5 million and MEP with $107.5 million (after a mandated 0.8% across the board rescission and a 0.54% rescission of Commerce, Justice, State discretionary accounts). The CREATE Act, P.L. 108-453, made changes in patent law to promote cooperative research among academia, government, and industry. P.L. 108-311 extended the research tax credit through December 31, 2005. In the 109th Congress, H.R. 250 establishes several new manufacturing technology programs for small and medium-sized firms. S. 296, introduced February 3, 2005, would authorize appropriations for MEP through FY2008, including $115 million in FY2006. The Administration's FY2006 budget request proposes $46.8 million for the Manufacturing Extension Partnership and no funding for the Advanced Technology Program.

BACKGROUND AND ANALYSIS
TECHNOLOGY AND COMPETITIVENESS

Interest in technology development and industrial innovation increased as concern mounted over the economic strength of the nation and over competition from abroad. For the United States to be competitive in the world economy, U.S. companies must be able to engage in trade, retain market shares, and offer high quality products, processes, and services while the nation maintains economic growth and a high standard of living. Technological advancement is important because the commercialization of inventions provides economic benefits from the sale of new products or services; from new ways to provide a service; or from new processes that increase productivity and efficiency. It is widely accepted that technological progress is responsible for up to one-half the growth of the U.S. economy, and is one principal driving force in long-term growth and increases in living standards.

Technological advances can further economic growth because they contribute to the creation of new goods, new services, new jobs, and new capital. The application of technology can improve productivity and the quality of products. It can expand the range of services that can be offered as well as extend the geographic distribution of these services. The development and use of technology also plays a major role in determining patterns of international trade by affecting the comparative advantages of industrial sectors. Since technological progress is not necessarily determined by economic conditions — it also can be influenced by advances in science, the organization and management of firms, government activity, or serendipity — it can have effects on trade independent of shifts in macroeconomic factors. New technologies also can help compensate for possible disadvantages in the cost of capital and labor faced by firms.

FEDERAL ROLE

In the recent past, American companies faced increased competitive pressures in the international marketplace from firms based in countries where governments actively promote commercial technological development and application. In the United States, the generation of technology for the commercial marketplace is primarily a private sector activity. The federal government traditionally becomes involved only for certain limited purposes. Typically these are activities which have been determined to be necessary for the "national good" but which cannot, or will not, be supported by industry.

To date, the U.S. government has funded research and development (R&D) to meet the mission requirements of the federal departments and agencies. It also finances efforts in areas where there is an identified need for research, primarily basic research, not being performed in the private sector. Federal support reflects a consensus that basic research is critical because it is the foundation for many new innovations. However, any returns created by this activity are generally long term, sometimes not marketable, and not always evident. Yet the rate of return to society as a whole generated by investments in research is significantly larger than the benefits that can be captured by the firm doing the work.

Many past government activities to increase basic research were based on a "linear" model of innovation. This theory viewed technological advancement as a series of sequential steps starting with idea origination and moving through basic research, applied research, development, commercialization, and diffusion into the economy. Increases in federal funds in the basic research stage were expected to result in concomitant increases in new products and processes. However, this linear concept is no longer considered valid. Innovations often occur that do not require basic or applied research or development; in fact most innovations are incremental improvements to existing products or processes. In certain areas, such as biotechnology, the distinctions between basic research and commercialization are small and shrinking. In others, the differentiation between basic and applied research is artificial. The critical factor is the *commercialization* of the technology. Economic benefits accrue only when a technology or technique is brought to the marketplace where it can be sold to generate income or applied to increase productivity. Yet, while the United States has a strong basic research enterprise, foreign firms appear more adept at taking the results of these scientific efforts and making commercially viable products. Often U.S. companies are competing in the global marketplace against goods and services developed by foreign industries from research performed in the United States. Thus, there has been increased congressional interest in mechanisms to accelerate the development and commercialization processes in the private sector.

The development of a governmental effort to facilitate technological advance has been particularly difficult because of the absence of a consensus on the need for an articulated policy. Technology demonstration and commercialization have traditionally been considered private sector functions in the United States. While over the years there have been various programs and policies (such as tax credits, technology transfer to industry, and patents), the approach had been ad hoc and uncoordinated. Much of the program development was based upon what individual committees judged appropriate for the agencies over which they have jurisdiction. Despite the importance of technology to the economy, technology-related considerations often have not been integrated into economic decisions.

There have been attempts to provide a central focus for governmental activity in technology matters. P.L. 100-519 created within the Department of Commerce a Technology Administration headed by a new Under Secretary for Technology. In November 1993, former President Clinton established a National Science and Technology Council to coordinate decisionmaking in science and technology and to insure their integration at all policy levels. However, technological issues and responsibilities remain shared among many departments and agencies. This diffused focus has sometimes resulted in actions which, if not at cross purposes, may not have accounted for the impact of policies or practices in one area on other parts of the process. Technology issues involve components which operate both separately and in concert. While a diffused approach can offer varied responses to varied issues, the importance of interrelationships may be underestimated and their usefulness may suffer.

Several times, Congress has examined the idea of an industrial policy to develop a coordinated approach on issues of economic growth and industrial competitiveness. Technological advance is both one aspect of this and an altogether separate consideration. In looking at the development of an identified policy for industrial competitiveness, advocates argue that such an effort could ameliorate much of the uncertainty with which the private sector perceives future government actions. It has been argued that consideration and delineation of national objectives could encourage industry to engage in more long-term planning with regard to R&D and to make decisions as to the best allocation of resources. Such a technology policy could generate greater consistency in government activities. Because technological development involves numerous risks, efforts to minimize uncertainty regarding federal programs and policies may help alleviate some of the disincentives perceived by industry.

The development of a technology policy, however, would require a new orientation by both the public and private sectors. There is widespread resistance to what could be and has been called national planning, due variously to doubts as to its efficacy, to fear of adverse effects on our market system, to political beliefs about government intervention in our economic system, and to the current emphasis on short- term returns in both the political and economic arenas. Yet proponents note that planning can be advisory or indicative rather than mandatory. The focus provided by a technology policy could arguably provide a more receptive or helpful governmental environment within which business can make better decisions. Advocates assert that it could also reassure industry of government's ongoing commitment to stimulating R&D and innovation in the private sector.

Consideration of what constitutes government policy (both in terms of the industrial policy and technology policy) covers a broad range of ideas from laissez-faire to special government incentives to target specific high-technology, high-growth industries. Suggestions have been made for the creation of federal mechanisms to identify and support strategic industries and technologies. Various federal agencies and private sector groups have developed critical technology lists. However, others maintain that such targeting is an unwanted, and unwarranted, interference in the private sector which will cause unnecessary dislocations in the marketplace or a misallocation of resources. The government does not have the knowledge or expertise to make business-related decisions. Instead, they argue, the appropriate role for government is to encourage innovative activities in all industries and to keep market related decisionmaking within the business community that has ultimate responsibility for commercialization and where such decisions have traditionally been made.

The relationship between government and industry is a major factor affecting innovation and the environment within which technological development takes place. This relationship often has been adversarial, with the government acting to regulate or restrain the business community, rather than to facilitate its positive contributions to the nation. However, the situation is changing; it has become increasingly apparent that lack of cooperation can be detrimental to the nation as it faces competition from companies in countries where close government-industry collaboration is the norm. There are an increasing number of areas where the traditional distinctions between public and private sector functions and responsibilities are becoming blurred. Many assumptions have been questioned, particularly in light of the increased internationalization of the U.S. economy. The business sector is no longer be viewed in an exclusively domestic context; the economy of the United States is often tied to the economies of other nations. The technological superiority long held by the United States in many areas has been challenged by other industrialized countries in which economic, social, and political policies and practices foster government-industry cooperation in technological development.

A major divergence from the past was evident in the approach taken by the former Clinton Administration. Articulated in two reports issued in February 1993 *(A Vision of Change for America and Technology for America's Economic Growth, A New Direction to Build Economic Strength)*, the proposal called for a national commitment to, and a strategy for, technological advancement as part of a defined national economic policy. This detailed strategy offered a policy agenda for economic growth in the United States, of which technological development and industrial competitiveness are critical components.

In articulating a national technology policy, the approach initially recommended and subsequently followed by the Administration was multifaceted and provided a wide range of options while for the most part reflecting current trends in congressional efforts to facilitate industrial advancement. This policy increased federal coordination and augmented direct government spending for technological development. While many past activities focused primarily on research, the new initiatives shifted the emphasis toward *development* of new products, processes, and services by the private sector for the commercial marketplace. In addition, a significant number of the proposals aimed to increase both government and private sector support for R&D leading to the commercialization of technology.

To facilitate technological advance, the Clinton approach focused on increasing investment; *investment* in research, primarily civilian research, to meet the Nation's needs in energy, environmental quality, and health; investment in the development and commercialization of new products, processes, and services for the marketplace; investment in improved manufacturing to make American goods less expensive and of better quality; investment in small, high technology businesses in light of their role in innovation and job creation; and investment in the country's infrastructure to support all these efforts. To make the most productive use of this increased investment, the Administration supported increased *cooperation* between all levels of government, industry, and academia to share risk, to share funding, and to utilize the strengths of each sector in reaching common goals of economic growth, productivity improvement, and maintenance of a high living standard. On November 23, 1993, President Clinton issued Executive Order 12881 establishing a National Science and Technology Council (NSTC), a cabinet-level body to "...coordinate science, space, and technology policies throughout the federal government."

The approach adopted by the former Administration has been questioned by recent Congresses and by the current Bush Administration. However, despite the continuing debate on what is the appropriate role of government and what constitutes a desirable government technology development policy, it remains an undisputed fact that what the government does or does not do affects the private sector and the marketplace. The various rules, regulations, and other activities of the government have become de facto policy as they relate to, and affect, innovation and technological advancement. It has been argued that these actions are not sufficiently understood or analyzed with respect to the larger context within which economic growth occurs. According to critics, these actions also are not coordinated in any meaningful way so that they promote an identifiable goal, whether that goal is as general as the "national welfare" or as specific as the growth of a particular industry.

LEGISLATIVE INITIATIVES AND CURRENT PROGRAMS

Legislative initiatives have reflected a trend toward expanding the government's role beyond traditional funding of mission-oriented R&D and basic research toward the facilitation of technological advancement to meet other critical national needs, including the economic growth that flows from new commercialization and use of technologies and techniques in the private sector. An overview of recent legislation shows federal efforts aimed at (1) encouraging industry to spend more on R&D; (2) assisting small high-technology businesses; (3) promoting joint research activities between companies; (4) fostering cooperative work between industry and universities; (5) facilitating the transfer of technology from the federal laboratories to the private sector; and (6) providing incentives for quality improvements. These efforts tend toward removing barriers to technology development in the private sector (thereby permitting market forces to operate) and providing incentives to encourage increased private sector R&D activities. While most focus primarily on research, some also involve policies and programs associated with technology development and commercialization.

Increased R&D Spending

To foster increased company spending on research, the 1981 Economic Recovery Tax Act (P.L. 97-34) mandated a temporary incremental tax credit for qualified research expenditures. The law provided a 25% tax credit for the increase in a firm's qualified research costs above the average expenditures for the previous three tax years. Qualified costs included in-house expenditures such as wages for researchers, material costs, and payments for use of equipment; 65% of corporate grants towards basic research at universities and other relevant institutions; and 65% of payments for contract research. The credit applied to research expenditures through 1985.

The Tax Reform Act of 1986 (P.L. 99-514) extended the research and experimentation (RandE) tax credit for another three years. However, the credit was lowered to 20% and is applicable to only 75% of a company's liability. The 1988 Tax Corrections Act (P.L. 100-647) approved a one-year extension of the research tax credit. The Omnibus Budget Reconciliation Act (P.L. 101-239) extended the credit through September 30, 1990 and made small start-up firms eligible for the credit. The FY1991 Budget Act (P.L. 101-508) again continued the tax credit provisions through 1992. The law expired in June 1992 when former President Bush vetoed H.R. 11 that year. However, P.L. 103-66, the Omnibus Budget Reconciliation Act of 1993, reinstated the credit through July 1995 and made it retroactive to the former expiration date. The tax credit again was allowed to expire until P.L. 104-188, the Small Business Job Protection Act, restored it from July 1, 1996 through May 31, 1997. P.L. 105-34, the Taxpayer Relief Act of 1997, extended the credit for 13 months from June 1, 1997 through June 30, 1998. Although it expired once again at the end of June, the Omnibus Consolidated Appropriations Act, P.L. 105-277, reinstated the tax credit through June 30, 1999. During the 105th Congress, various bills were introduced to make the tax credit permanent; other bills would have allowed the credit to be applied to certain collaborative research consortia. On August 5, 1999, both the House and Senate agreed to the conference report for H.R. 2488, the Financial Freedom Act, which would have extended the credit for five years through June 30, 2004. This bill also would have increased the credit rate applicable under the alternative incremental research credit by one percentage point per step. While the President vetoed this bill on September 23, 1999, the same provisions are included in Title V of P.L. 106-170 signed into law on December 17, 1999. P.L. 108-311 extends the research tax credit through December 31, 2005.

The Small Business Development Act (P.L. 97-219), as extended (P.L. 99-443), established a program to facilitate increased R&D within the small-business, high-technology community. Each federal agency with a research budget was required to set aside 1.25% of its R&D funding for grants to small firms for research in areas of interest to that agency. P.L. 102-564, which reauthorized the SBIR program, increased the set-aside to 2.5%, phased in over a five-year period. Funding is, in part, dependent on companies obtaining private sector support for the commercialization of the resulting products or processes. The authorization for the program was set to terminate October 1, 2000. However, the SBIR activity was reauthorized through September 30, 2008 by P.L. 106-554, signed into law on December 21, 2000. A pilot effort, the Small Business Technology Transfer (STTR) program, also was created to encourage firms to work with universities or federal laboratories to commercialize the results of research. This program is funded by a 0.15% (phased in) set-aside. Set to expire in FY1997, the STTR originally was extended for one year until P.L. 105-135 reauthorized this activity through FY2001. Passed in the current Congress, P.L. 107-50 extends the program through FY2009 and expands the set-aside to 0.3% beginning in FY2004. Also in FY2004, the amount of individual Phase II grants increases to $750,000. (See CRS Report 96-402, *Small Business Innovation Research Program*.)

The Omnibus Trade and Competitiveness Act of 1988 (P.L. 100-418) created the Advanced Technology Program (ATP) at the Department of Commerce's National Institute of Standards and Technology. ATP provides seed funding, matched by private sector investment, for companies or consortia of universities, industries, and/or government laboratories to accelerate development of generic technologies with broad application across industries. The first awards were made in 1991. As of May 2004, 736 projects have been funded representing approximately $2.2 billion in federal dollars matched by $2 billion in private sector financing. About 66% of the awardees are small businesses or cooperative efforts led by such firms. (For more information, see CRS Report 95-36, *The Advanced Technology Program*.)

Appropriations for the ATP include $35.9 million in FY1991, $47.9 million in FY1992, and $67.9 million in FY1993. FY1994 appropriations increased significantly to $199.5 million and even further in FY1995 to $431 million. However, P.L. 104-6, rescinded $90 million from this amount. There was no FY1996 authorization. The original FY1996 appropriations bill, H.R. 2076, which passed the Congress was vetoed by the President, in part, because it provided no support for ATP. The appropriations legislation finally enacted, P.L. 104-134, did fund the Advanced Technology Program at $221 million. For FY1997, the President's budget request was $345 million. Again, there was no authorizing legislation. However, P.L. 104-208, the Omnibus Consolidated Appropriations Act, provided $225 million for ATP, later reduced by $7 million to $218 million by P.L. 105-18, the FY1997 Emergency Supplemental Appropriations and Rescission Act. For FY1998, the Administration requested $276 million in funding. P.L. 105-119, appropriated FY1998 financing of ATP at $192.5 million, again at a level less than the previous year. The Administration's FY1999 budget proposal included $259.9 million for this program, a 35% increase. While not providing such a large increase, P.L. 105-277 did fund ATP for FY1999 at $197.5 million, 3% above the previous year. This figure reflected a $6 million rescission contained in the same law that accounted for "deobligated" funds resulting from early termination of certain projects.

In FY2000, the President requested $238.7 million for ATP, an increase of 21% over the previous year. S. 1217, as passed by the Senate, would have appropriated $226.5 million for ATP. H.R. 2670, as passed by the House, provided no funding for the activity. The report to accompany the House bill stated that there was insufficient evidence ". . . to overcome those fundamental questions about whether the program should exist in the first place." Yet, P.L. 106-113 eventually did finance the program at $142.6 million, 28% below prior year funding. The Clinton Administration's FY2001 budget included $175.5 million for the Advanced Technology Program, an increase of 23% over the earlier fiscal year. Once again, the original version of the appropriations bills that passed the House did not contain any financial support for the activity. However, P.L. 106-553 provided $145.7 million in FY2001 support for ATP, 2% above the previous funding level.

For FY2002, President Bush's budget proposed suspending all funding for new ATP awards pending an evaluation of the program. In the interim, $13 million would have been provided to meet the financial commitments for on-going projects. H.R. 2500, as initially passed by the House, also did not fund new ATP grants but offered $13 million for prior commitments. The version of H.R. 2500 that originally passed the Senate provided $204.2 million for the ATP effort. P.L. 107-77 funds the program at $184.5 million, an increase of almost 27% over the previous fiscal year.

The Administration's FY2003 budget request would have funded the Advanced Technology Program at $108 million; 35% below the FY2002 appropriation level. No relevant appropriations legislation was passed by the 107th Congress; a series of Continuing Resolutions funded the program until the 108th Congress enacted P.L. 108-7 which financed ATP at $178.8 million for FY2003 (after the mandated 0.65% across the board recision). H.R. 175 would abolish the program.

In its FY2004 budget, the Administration proposed to provide $17 million to cover ongoing commitments to ATP; however no new projects would be funded. H.R. 2799, the FY2004 appropriations bill first passed by the House on July 23, 2003, provided no support for ATP. Subsequently incorporated into H.R. 2673, which became P.L. 108-199, the FY2004 Consolidated Appropriations Act, the new legislation funds ATP at $179.2 million (prior to a mandated 0.59% across the board rescission). As reported to the Senate from the Committee on Appropriations, S. 1585 would have financed the program at $259.6 million.

The President's FY2005 budget, as well as H.R. 4754, the Commerce, Justice, State Appropriations bill originally passed by the House, did not include any funding for ATP. As reported to the Senate from the Committee on Appropriations, S. 2809 would have provided $203 million for the program, 19% above the previous fiscal year. P.L. 108-447, the FY2005 Omnibus Appropriations Act, funds ATP at $136.5 million (after several rescissions mandated in the legislation), 20% below FY2004.

For FY2006, the Administration's budget does not include funding for the Advanced Technology Program.

Industry-University Cooperative Efforts

The promotion of cooperative efforts among academia and industry is aimed at increasing the potential for the commercialization of technology. (For more information, see CRS Issue Brief IB89056, *Cooperative R&D: Federal Efforts to Promote Industrial Competitiveness*.) Traditionally, basic research has been performed in universities or in the federal laboratory system while the business community focuses on the manufacture or provision of products, processes, or services. Universities are especially suited to undertake basic research. Their mission is to educate and basic research is an integral part of the educational process. Universities generally are able to undertake these activities because they do not have to produce goods for the marketplace and therefore can do research not necessarily tied to the development of a commercial product or process.

Subsequent to World War II, the federal government supplanted industry as the primary source of funding for basic research in universities. It also became the principal determinant of the type and direction of the research performed in academia. This resulted in a disconnect between the university and industrial communities. The separation and isolation of the parties involved in the innovation process is thought to be a barrier to technological progress. The difficulties in moving an idea from the concept stage to a commercial product or process are compounded when several entities are involved. Legislation to stimulate cooperative efforts among those involved in technology development is viewed as one way to promote innovation and facilitate the international competitiveness of U.S. industry.

Several laws have attempted to encourage industry-university cooperation. Title II of the Economic Recovery Tax Act of 1981 (P.L. 97-34) provided, in part, a 25% tax credit for 65% of all company payments to universities for the performance of basic research. Firms were also permitted a larger tax deduction for charitable contributions of equipment used in scientific research at academic institutions. The Tax Reform Act of 1986 (P.L. 99-514) kept this latter provision, but reduced the credit for university basic research to 20% of all corporate expenditures for this over the sum of a fixed research floor plus any decrease in non-research giving.

The 1981 Act also provided an increased charitable deduction for donations of new equipment by a manufacturer to an institution of higher education. This equipment must be used for research or research training for physical or biological sciences within the United States. The tax deduction is equal to the manufacturer's cost plus one-half the difference between the manufacturer's cost and the market value, as long as it does not exceed twice the cost basis. These provisions were extended through July 1995 by the Omnibus Budget Reconciliation Act of 1993, but then expired until restored by the passage of P.L. 104-188, P.L. 105-277, and P.L. 106-170 as noted above.

Amendments to the patent and trademark laws contained in P.L. 96-517 (commonly called the "Bayh-Dole Act") also were designed to foster interaction between academia and the business community. This law provides, in part, for title to inventions made by contractors receiving federal R&D funds to be vested in the contractor if they are small businesses, universities, or not-for-profit institutions. Certain rights to the patent are reserved for the government and these organizations are required to commercialize within a predetermined and agreed upon time frame. Providing universities with patent title is expected to encourage licensing to industry where the technology can be manufactured or used thereby creating a financial return to the academic institution. University patent applications and licensing have increased significantly since this law was enacted. (See CRS Report RL32076, *The Bayh-Dole Act: Selected Issues in Patent Policy and the Commercialization of Technology*, CRS Report RL30320, *Patent Ownership and Federal Research and Development: A Discussion on the Bayh-Dole Act and the Stevenson-Wydler Act*, and CRS Report 98-862, *R&D Partnerships and Intellectual Property: Implications for U.S. Policy*.)

The CREATE Act, P.L. 108-453, makes changes in the patent laws to promote cooperative research and development among universities, government, and the private sector. The bill amends section 103(c) of title 25, United States Code, such that certain actions between researchers under a joint research agreement will not preclude patentability.

Joint Industrial Research

Private sector investments in basic research are often costly, long term, and risky. Although not all advances in technology are the result of research, it is often the foundation of important new innovations. To encourage increased industrial involvement in research, legislation was enacted to allow for joint ventures in this arena. It is argued that cooperative research reduces risks and costs and allows for work to be performed that crosses traditional boundaries or expertise and experience. Such collaborative efforts make use of existing and support the development of new resources, facilities, knowledge, and skills.

The National Cooperative Research Act (P.L. 98-462) encourages companies to undertake joint research. The legislation clarifies the antitrust laws and requires that a "rule of reason" standard be applied in determinations of violations of these laws; cooperative research ventures are not to be judged illegal "per se." It eliminates treble damage awards for those research ventures found in violation of the antitrust laws if prior disclosure (as defined in the law) has been made. P.L. 98-462 also makes changes in the way attorney fees are awarded. Defendants can collect attorney fees in specified circumstances, including when the claim is judged frivolous, unreasonable, without foundation, or made in bad faith. However, the attorney fee award to the prevailing party may be offset if the court decides that the prevailing party conducted a portion of the litigation in a manner which was frivolous, unreasonable, without foundation, or in bad faith. These provisions were included to discourage frivolous litigation against joint research ventures without simultaneously discouraging suits of plaintiffs with valid claims. Over 700 joint research ventures have filed with the Department of Justice since passage of this legislation.

P.L. 103-42, the National Cooperative Production Amendments Act of 1993, amends the National Cooperative Research Act by, among other things, extending the original law's provisions to joint manufacturing ventures. These provisions are only applicable, however, to cooperative production when (1) the principal manufacturing facilities are "...located in the United States or its territories, and (2) each person who controls any party to such venture...is a United States person, or a foreign person from a country whose law accords antitrust treatment no less favorable to United States persons than to such country's domestic persons with respect to participation in joint ventures for production."

Commercialization of the Results of Federally Funded R&D

Another approach to encouraging the commercialization of technology involves the transfer of technology from federal laboratories and contractors to the private sector where commercialization can proceed. Because the federal laboratory system has extensive science and technology resources and expertise developed in pursuit of mission responsibilities, it is a potential source of new ideas and knowledge which may be used in the business community. (See CRS Issue Brief IB85031, *Technology Transfer: Utilization of Federally Funded Research and Development*, for more details.)

Despite the potential offered by the resources of the federal laboratory system, however, the commercialization level of the results of federally funded R&D remained low. Studies indicated that only approximately 10% of federally owned patents were ever utilized. There are many reasons for this low level of usage, one of which is the fact that some technologies and/or patents have no market application. However, industry unfamiliarity with these technologies, the "not-invented-here" syndrome, and perhaps more significantly, the ambiguities associated with obtaining title to or exclusive license to federally owned patents also contribute to the low level of commercialization.

Over the years, several governmental efforts have been undertaken to augment industry's awareness of federal R&D resources. The Federal Laboratory Consortium for Technology Transfer was created in 1972 (from a Department of Defense program) to assist in transferring technology from the federal government to state and local governments and the private sector. To expand on the work of the Federal Laboratory Consortium, and to provide added emphasis on the commercialization of government technology, Congress passed P.L. 96-480, the Stevenson-Wydler Technology Innovation Act of 1980. Prior to this law, technology transfer was not an explicit mandate of the federal departments and agencies with the exception of the National Aeronautics and Space Administration. To provide "legitimacy" to the numerous technology activities of the government, Congress, with strong bipartisan support, enacted P.L. 96-480 which explicitly states that the federal government has the responsibility, "...to ensure the full use of the results of the nation's federal investment in research and development." Section 11 of the law created a system within the federal government to identify and disseminate information and expertise on what technologies or techniques are available for transfer. Offices of Research and Technology Applications were established in each federal laboratory to distinguish technologies and ideas with potential applications in other settings.

Several amendments to the Stevenson-Wydler Technology Innovation Act have been enacted to provide additional incentives for the commercialization of technology. P.L. 99-502, the Federal Technology Transfer Act, authorizes activities designed to encourage industry, universities, and federal laboratories to work cooperatively. It also establishes incentives for federal laboratory employees to promote the commercialization of the results of federally funded research and development. The law amends P.L. 96-480 to allow government-owned, government-operated laboratories to enter into cooperative R&D agreements (CRADAs) with universities and the private sector. This authority is extended to government-owned, contractor-operated laboratories by the Department of Defense FY1990 Authorization Act, P.L. 101-189. (See CRS Report 95-150, *Cooperative Research and Development Agreements (CRADAs)*.) Companies, regardless of size, are allowed to retain title to inventions resulting from research performed under cooperative agreements. The federal government retains a royalty-free license to use these patents. The Technology Transfer Improvements and Advancement Act (P.L. 104- 113), clarifies the dispensation of intellectual property rights under CRADAs to facilitate the implementation of these cooperative efforts. The Federal Laboratory Consortium is given a legislative mandate to assist in the coordination of technology transfer. To further promote the use of the results of federal R&D, certain agencies are mandated to create a cash awards program and a royalty sharing activity for federal scientists, engineers, and technicians in recognition of efforts toward commercialization of this federally developed technology. These efforts are facilitated by a provision of the National Defense Authorization Act for FY1991 (P.L. 101-510), which amends the Stevenson-Wydler Technology Innovation Act to allow government agencies and laboratories to develop partnership intermediary programs to augment the transfer of laboratory technology to the small business sector.

Amendments to the Patent and Trademark law contained in Title V of P.L. 98- 620 make changes which are designed to improve the transfer of technology from the federal laboratories — especially those operated by contractors — to the private sector and increase the chances of successful commercialization of these technologies. This law permits the contractor at government-owned, contractor-operated laboratories (GOCOs) to make decisions at the laboratory level as to the granting of licenses for subject inventions. This has the potential of effecting greater interaction between laboratories and industry in the transfer of technology. Royalties on these inventions are also permitted to go back to the contractor to be used for additional R&D, awards to individual inventors, or education. While there is a cap on the amount of the royalty returning directly to the lab in order not to disrupt the agency's mission requirements and congressionally mandated R&D agenda, the establishment of discretionary funds gives contractor-operated laboratories added incentive to encourage technology transfer.

Under P.L. 98-620, private companies, regardless of size, are allowed to obtain exclusive licenses for the life of the patent. Prior restrictions allowed large firms use of exclusive license for only 5 of the 17 years (now 20 years) of the life of the patent. This should encourage improved technology transfer from the federal laboratories or the universities (in the case of university operated GOCOs) to large corporations which often have the resources necessary for development and commercialization activities. In addition, the law permits GOCOs (those operated by universities or nonprofit institutions) to retain title to inventions made in the laboratory within certain defined limitations. Those laboratories operated by large companies are not included in this provision.

P.L. 106-404, the Technology Transfer Commercialization Act, alters current practices concerning patents held by the government to make it easier for federal agencies to license such inventions. The law amends the Stevenson-Wydler Technology Innovation Act and the Bayh-Dole Act to decrease the time delays associated with obtaining an exclusive or partially exclusive license. Previously, agencies were required to publicize the availability of technologies for three months using the *Federal Register* and then provide an additional 60 day notice of intent to license by an interested company. Under the new legislation, the time period is shorten to 15 days in recognition of the ability of the internet to offer widespread notification and the necessity of time constraints faced by industry in commercialization activities. Certain rights are retained by the government. The bill also allows licenses for existing government-owned inventions to be included in CRADAs.

The Omnibus Trade and Competitiveness Act (P.L. 100-418) mandated the creation of a program of regional centers to assist small manufacturing companies to use knowledge and technology developed under the auspices of the National Institute of Standards and Technology and other federal agencies. Federal funding for the centers is matched by non-federal sources including state and local governments and industry. Originally, seven Regional Centers for the Transfer of Manufacturing Technology were selected. Later, the initial program was expanded in 1994 to create the Manufacturing Extension Partnership (MEP) to meet new and growing needs of the community. In a more varied approach, the Partnership involves both large centers and smaller, more dispersed organizations sometimes affiliated with larger centers as well as the NIST State Technology Extension Program which provides states with grants to develop the infrastructure necessary to transfer technology from the federal government to the private sector (an effort which was also mandated by P.L. 100-418) and a program which electronically ties the disparate parties together along with other federal, state, local, and academic technology transfer organizations. There are now centers in all 50 states and Puerto Rico. Since the manufacturing extension activity was created in 1989, awards made by NIST have resulted in the creation of approximately 350 regional offices. [It should be noted that the Department of Defense also funded 36 centers through its Technology Reinvestment Project (TRP) in FY1994 and FY1995. When the TRP was terminated, NIST took over support for 20 of these programs in FY1996 and funded the remaining efforts during FY1997.]

Funding for this program was $11.9 million in FY1991, $15.1 million in FY1992, and $16.9 million in FY1993. In FY1994 support for the expanded Manufacturing Technology Partnerships was $30.3 million. The following fiscal year, P.L. 103-317 appropriated $90.6 million for this effort, although P.L. 104-19 rescinded $16.3 million from this amount. While the original FY1996 appropriations bill, H.R. 2076, was vetoed by the President, the $80 million funding for MEP was retained in the final legislation, P.L. 104-134. The President's FY1997 budget request was $105 million. No FY1997 authorization legislation was enacted, but P.L. 104-208 appropriated $95 million for Manufacturing Extension while temporarily lifting the six-year limit on federal support for individual centers. The Administration requested FY1998 funding of $123 million. Again no authorizations were passed. However, the FY1998 appropriations bill, P.L. 105-119, financed the MEP program at $113.5 million. This law also permitted government funding, at one-third the centers total annual cost, to continue for additional periods of one year over the original six-year limit, if a positive evaluation is received. The President's FY1999 budget included $106.8 million for the MEP, a 6% decrease from current funding. The Omnibus Consolidated Appropriates Act, P.L. 105-277, appropriated the $106.8 million. The decrease in funding reflects a reduced federal financial commitment as the centers mature, not a decrease in program support. In addition, the Technology Administration Act of 1998, P.L. 105-309, permits the federal government to fund centers at one-third the cost after the six years if a positive, independent evaluation is made every two years.

For FY2000, the Administration requested $99.8 million in support for the MEP. Again, the lower federal share indicated a smaller statutory portion required of the government. S. 1217, as passed by the Senate, would have appropriated $109.8 million for the Manufacturing Extension Partnership, an increase of 3% over FY1999. H.R. 2670, as passed initially by the House, would have appropriated $99.8 million for this activity. The version of the H.R. 2670 passed by both House and Senate provided FY2000 appropriations of $104.8 million. While the President vetoed that bill, the legislation that was ultimately enacted, P.L. 106-113, appropriated $104.2 million after the mandated rescission. The Clinton Administration's FY2001 budget requested $114.1 million for the Partnership, an increase of almost 9% over current support. Included in this figure was funding to allow the centers to work with the Department of Agriculture and the Small Business Administration on an e-commerce outreach program. P.L. 106-553 appropriates $105.1 million for FY2001, but does not fund any new initiatives.

The FY2002 Bush Administration budget proposed providing $106.3 million for MEP. H.R. 2500, as originally passed by the House, would have funded MEP at $106.5 million. The initial version of H.R. 2500 passed by the Senate would have provided $105.1 million for the program. The final legislation, P.L. 107-77 funds the Partnership at $106.5 million.

For FY2003, the Administration's budget included an 89% decrease in support for MEP. According to the budget document, "...consistent with the program's original design, the President's budget recommends that all centers with more than six years experience operate without federal contribution." A number of Continuing Resolutions supported the Partnership at FY2002 levels until the 108th Congress enacted P.L. 108-7 which appropriates $105.9 million for MEP in FY2003 (after the mandated recision).

The President's FY2004 budget requested $12.6 million for MEP to finance only those centers that have not reached six years of federal support. H.R. 2799, as initially passed by the House, would have appropriated $39.6 million for the Partnership. This bill was subsequently incorporated into H.R. 2673, which became P.L. 108-199, the FY2004 Consolidated Appropriations Act. This legislation finances MEP at $38.7 million after a mandated rescission. S. 1585, reported to the Senate by the Committee on Appropriations, would have funded the program at $106.6 million. (For additional information see CRS Report 97-104, *Manufacturing Extension Partnership Program: An Overview*.)

The Administration proposed funding MEP at $39.2 million for FY2005. H.R. 4754, as originally passed by the House on July 8, 2004, would have appropriated $106 million for this program. S. 2809, as reported by the Senate Committee on Appropriations would have provided $112 million for MEP to "fully fund" existing centers and provide assistance to small and rural States. P.L. 108-447, the FY2005 Omnibus Appropriations Act, funds manufacturing extension at $107.5 million (after several mandated rescissions included in the legislation).

For FY2006, the President's budget requests $46.8 million for the Manufacturing Extension Partnership, 56% below funding for the current fiscal year.

DIFFERENT APPROACH?

As indicated above, the laws affecting the R&D environment have included both direct and indirect measures to facilitate technological innovation. In general, direct measures are those which involve budget outlays and the provision of services by government agencies. Indirect measures include financial incentives and legal changes (e.g., liability or regulatory reform; new antitrust arrangements). Supporters of indirect approaches argue that the market is superior to government in deciding which technologies are worthy of investment. Mechanisms that enhance the market's opportunities and abilities to make such choices are preferred. Advocates further state that dependency on agency discretion to assist one technology in preference to another will inevitably be subjected to political pressures from entrenched interests. Proponents of direct government assistance maintain, conversely, that indirect methods can be wasteful and ineffective and that they can compromise other goals of public policy in the hope of stimulating innovative performance. Advocates of direct approaches argue that it is important to put the country's scarce resources to work on those technologies that have the greatest promise as determined by industry and supported by its willingness to match federal funding.

In the past, while Republicans tended to prefer reliance on free market investment, competition, and indirect support by government, participants in the debates generally did not make definite (or exclusionary) choices between the two approaches, nor consistently favor one over the other. For example, some proponents of a stronger direct role for the government in innovation are also supporters of enhanced tax preferences for R&D spending, an indirect mechanism. Opponents of direct federal support for specific projects (e.g., SEMATECH, flat panel displays) may nevertheless back similar activities focused on more general areas such as manufacturing or information technology. However, the 104th Congress directed their efforts at eliminating or curtailing many of the efforts which previously had enjoyed bipartisan support. Initiatives to terminate the Advanced Technology Program, funding for flat panel displays, and agricultural extension reflected concern about the role of government in developing commercial technologies. The Republican leadership stated that the government should directly support basic science while leaving technology development to the private sector. Instead of federal funding, changes to the tax laws, proponents argue, will provide the capital resources and incentives necessary for industry to further invest in R&D. Many of the same issues were considered in subsequent congresses. While funding for several programs decreased, support for most on-going activities continued, some at increased levels. How the debate over federal funding evolves in the 109th Congress may serve to redefine thinking about the government's efforts in promoting technological advancement in the private sector.

LEGISLATION IN THE 108TH CONGRESS

P.L. 108-7, H.J.Res. 2

Omnibus FY2003 Appropriations Act. Among other things, funds the Advanced Technology Program at $180 million and the Manufacturing Extension Partnership at $106.6 million. Introduced January 7, 2003; referred to the House Committee on Appropriations. Passed House on January 8, 2003. Passed Senate, amended, on January 23, 2003. House and Senate agreed to conference report on February 13, 2003. Signed into law by the president on February 20, 2003.

P.L. 108-199, H.R. 2673

FY2004 Consolidated Appropriations Act. Among other things, funds the Advanced Technology Program at $179.2 million in FY2004 and the Manufacturing Extension Partnership at $39.6 million (prior to a 0.59% across the board rescission). H.R. 2673 reported to the House as an original measure from the House Committee on Appropriations July 9, 2003. Passed the House July 14, 2003 and passed the Senate November 6, 2003. House agreed to conference report on December 8, 2003. Senate agreed to conference report on January 22, 2004. Signed into law by the President on January 23, 2004.

P.L. 108-311, H.R. 1308

Amends the Internal Revenue Code of 1986 to extend the research tax credit through December 31, 2005, among other things. Introduced March 18, 2004; referred to the House Committee on Ways and Means. Passed House March 19, 2004 and passed the Senate with an amendment on June 5, 2004. Both the House and the Senate agreed to the conference report on September 23, 2004. Signed into law by the President on October 4, 2004.

P.L. 108-447, H.R. 4818

FY2005 Omnibus Appropriations Act. Among other things, funds ATP at $136.5 million in FY2005 and MEP at $107.5 million after several rescissions mandated in the legislation. Introduced July 13, 2004; referred to the House Committee on Appropriations. Passed the House on July 15, 2004. Passed the Senate, amended, on September 23, 2004. Conference report agreed to in both the House and Senate on November 20, 2004. Signed into law on December 8, 2004.

P.L. 108-453, S. 2192

The CREATE Act. Amends patent law to promote cooperative research among universities, government, and the private sector. Introduced March 10, 2004; referred to the Senate Committee on the Judiciary. Reported to the Senate on April 29, 2004. Passed Senate on June 25, 2004. Passed House on November 20, 2004. Signed into law on December 10, 2004.

LEGISLATION

H.R. 250 (Ehlers)

Manufacturing Technology Competitiveness Act. Creates an interagency committee to coordinate federal manufacturing R&D. Establishes several new pilot grant programs in collaborative manufacturing research, among other things. Introduced January 6, 2005; referred to the Committee on Science.

S. 296 (Kohl)

Authorizes appropriations for the Manufacturing Extension Partnership through FY2008, including $115 million for FY2006. Introduced February 3, 2005; referred to the Senate Committee on Commerce, Science, and Transportation.

Chapter 4

SAFE HARBOR FOR PRECLINICAL USE OF PATENTED INVENTIONS IN DRUG RESEARCH AND DEVELOPMENT: *MERCK KGAA V. INTEGRA LIFESCIENCES I, LTD.*[*]

Brian T. Yeh

ABSTRACT

In *Merck KGaA v. Integra Lifesciences I, Ltd.*, __ U.S. __, 125 S. Ct. 2372 (2005), the United States Supreme Court unanimously held that the preclinical use of patented inventions in drug research is exempted from patent infringement claims by the "safe harbor" provision of the Patent Act, 35 U.S.C. § 271(e)(1). (Merck KGaA is a German company unaffiliated with the U.S.-based pharmaceutical company Merck and Co.) This decision potentially may help expedite the development of new medical treatments and lower the cost of some drugs for consumers.

In 2003, the U.S. Court of Appeals for the Federal Circuit had narrowly construed the safe harbor provision as protecting only clinical research activities that produce information for submission to the Food and Drug Administration (FDA) in the regulatory process. In vacating that decision, the U.S. Supreme Court ruled that the exemption applies to *all* uses of patented inventions that are "reasonably related" to the process of developing any information for FDA submission. The Court explained that, under certain conditions, the safe harbor provision is even "sufficiently broad" to protect the use of patented compounds in experiments that are not ultimately submitted to the FDA or drug experiments that are not ultimately the subject of an FDA submission. Finally, the scope of the exemption is not limited only to preclinical studies pertaining to a drug's safety in humans, but also includes preclinical data regarding a drug's efficacy, mechanism of action, pharmacokinetics, and pharmacology.

However, the Court cautioned that the exemption does not reach all experimental activity that at some point, however attenuated, may lead to an FDA approval process. For example, the safe harbor provision does not embrace basic scientific research

---

[*] Excerpted from CRS Report RL33114, dated October 7, 2005.

performed on a patented compound without the intent to develop a particular drug or without a reasonable belief that the compound will cause a particular physiological effect that the researcher desires. In addition, because the matter was not at issue in the case, the Court expressly declined to decide whether or to what extent the exemption applies to patented "research tools" that are often used to facilitate general research in developing compounds for FDA submissions.

INTRODUCTION

In *Merck KGaA v. Integra Lifesciences I, Ltd.*, __ U.S. __, 125 S. Ct. 2372 (2005), the United States Supreme Court decided, without dissent, that the patent law's safe harbor provision exempts from infringement the preclinical use of patented inventions in drug research. Without this legal immunity, pharmaceutical companies face patent infringement liability when they conduct preclinical experiments using rival companies' patented compounds. The U.S. Court of Appeals for the Federal Circuit had earlier found that the statutory exemption applied only to clinical research activity that contributes "relatively directly" to information the Food and Drug Administration (FDA) considers in approving a drug.[1] This narrow interpretation of the safe harbor provision had raised concerns that the patent law could significantly restrict the development and introduction of new medical treatments and generic drugs.

Vacating the appellate court's decision, the U.S. Supreme Court unanimously ruled that the exemption protects all uses of patented inventions that are "reasonably related" to the process of developing any information for FDA submission, which includes preclinical studies. The Court's expansive construction of the safe harbor provision "leaves adequate space for experimentation and failure on the road to regulatory approval"[2] and "provides a wide berth for the use of patented drugs in activities related to the federal regulatory process."[3]

BACKGROUND

It is normally a violation of the Patent Act to use any patented invention without prior authorization of the patent owner.[4] However, a statutory exception to this general rule provides: "It shall not be an act of infringement to make, use, offer to sell, or sell within the United States or import into the United States a patented invention ... solely for uses reasonably related to the development and submission of information" to the United States Food and Drug Administration (FDA).[5] Thus, a party that uses a patented invention without the patent owner's permission is committing an infringing act, but if the use comes within the scope of the statutory exception, the party will not be held liable for violating the patent owner's rights.

The Hatch-Waxman Act

The statutory exception was created by the Drug Price Competition and Patent Term Restoration Act of 1984,[6] commonly known as the Hatch-Waxman Act. This legislation modified the Patent Act by creating a new section, 35 U.S.C. § 271(e), that provides "safe harbor" from infringement for pharmaceutical companies using patented inventions in their drug research and development operations.

The Hatch-Waxman Act is widely credited with encouraging and expediting the creation and availability of generic versions of approved patented drugs. Prior to its enactment, pharmaceutical companies had to wait until all relevant patents expired before undertaking the clinical research necessary to obtain FDA approval of generic equivalents.[7] Thus, an established drug's patent term was *de facto* extended beyond its expiration date by the length of the FDA regulatory process for approving the generic equivalent, which took more than two years.[8] The Hatch-Waxman Act allows generic drug manufacturers to conduct safety and effectiveness tests during the time the brand name drug's patent is still in force, often resulting in immediate introduction of a generic drug into the market upon the pioneer drug's patent expiration.[9]

The FDA Drug Approval Process

The Federal Food, Drug, and Cosmetics Act (FDCA) regulates the manufacture, use, or sale of drugs.[10] Under the FDCA, the FDA must determine that a drug is safe and effective before it can be marketed to consumers. The FDCA establishes a two-stage approval process for new drugs: an "Investigational New Drug" (IND) application and a "New Drug Application" (NDA).[11]

The drug manufacturer must file an IND with the FDA after the company has identified, through preclinical testing on animals and in test tubes, chemical compounds that appear to have beneficial therapeutic effects. The IND is a request for authorization to conduct clinical (human) testing, and it must contain information and data from the preclinical studies that justify the proposed clinical trial.[12]

Once the FDA approves the IND, the drugmaker can commence clinical studies. If these studies demonstrate that a new drug is reasonably safe and effective for use, the drugmaker is required to submit a NDA.[13] The NDA must include data from preclinical and clinical studies. After extensive review of the NDA, the FDA issues final approval or denial of the application for manufacturing and selling the new drug to the public.[14]

THE SCOPE OF SAFE HARBOR

The Patent Act's safe harbor provision has often been compared to the "fair use" defense in copyright law, since it immunizes from liability otherwise infringing acts in order to advance compelling public policy interests. The legislative history of the Hatch-Waxman Act provides the basis for this analogy: "Just as we have recognized the doctrine of fair use in copyright, it is appropriate to create a similar mechanism in the patent law. That is all this bill does."[15] Despite this deceptively simple language of purpose, the safe harbor provision has been the subject of confusion and litigation for many years following its enactment. For over two decades, federal courts struggled to define the breadth and contours of the exemption, particularly concerning the types and uses of patented invention covered by the safe harbor.

Types Covered

As for the types of covered patented invention, the United States Supreme Court in *Eli Lilly and Co. v. Medtronic, Inc.* expansively interpreted § 271(e)(1) to include not only drug and veterinary products, but also medical devices that are subject to pre-market approval by the FDA.[16] The *Eli Lilly* Court determined that "[t]he phrase 'patented invention' in § 271(e)(1) is defined to include all inventions, not drug-related inventions alone."[17] The Court opined that if Congress had wanted the safe harbor to cover only generic drugs, "there were available such infinitely more clear and simply ways of expressing that intent."[18] As written, § 271(e)(1) applies to the "entire statutory scheme of regulation,"[19] including "medical devices, food additives, color additives, new drugs, antibiotic drugs, and human biological products."[20]

Uses Covered

Concerning the protected uses of a patented invention, a long disputed issue was what kind of research in the drug development process qualified for the exemption: basic research, preclinical research, or clinical studies. These three stages of drug development are described as follows: basic research involves the testing of thousands of compounds to discover any biological activity relevant to understanding the cause of a disease; the preclinical stage involves more focused research on a smaller group of chemical compounds in the hopes of finding the best candidate for clinical development; and clinical studies are the testing of the drug on human subjects in preparation for FDA approval.[21] Following its interpretive lead in *Eli Lilly*, the Supreme Court in *Merck KGaA v. Integra Lifesciences I, Ltd.("Integra")* ruled that § 271(e)(1) immunizes from infringement both preclinical and clinical use of patented inventions in the drug research and development process.

MERCK KGAA V. INTEGRA LIFESCIENCES I, LTD.

Facts of the Case

Integra Lifesciences I, Ltd. ("Integra") is a pharmaceutical company that owns five patents related to a sequence of three amino acids, arginine, glycine, and aspartic acid (the "RGD peptide"), which promotes cell adhesion by attaching to receptors on cell surface proteins called integrins.[22] Scientists working for Telios Pharmaceuticals, Inc. discovered that the RGD peptide had potential use in promoting wound healing and biocompatibility of prosthetic devices, prompting Telios to obtain patents for the RGD peptide compositions and methods. However, after failing to develop a viable commercial product, Telios sold the patents to Integra.[23]

In the mid-1980s, Dr. David Cheresh at the Scripps Research Institute ("Scripps"), a non-profit corporation that conducts biochemical research, discovered that blocking integrin receptors using the RGD peptide inhibited angiogenesis, a process by which new blood vessels sprout from existing vessels. Angiogenesis plays a critical role in the spread of many diseases, including cancerous tumor growth, diabetic retinopathy, and rheumatoid arthritis.[24]

Merck KGaA ("Merck"),[25] a German pharmaceutical corporation unaffiliated with the U.S.-based pharmaceutical company Merck and Co., was interested in developing this discovery into a drug to control angiogenesis. In 1988, Merck entered into an agreement with Scripps to provide funding for Dr. Cheresh's research, in exchange for Scripps granting Merck an option to license future discoveries arising from his research.[26] In 1994, Dr. Cheresh succeeded in reversing tumor growth in chicken embryos using a RGD peptide identified as EMD 66203, which had been provided by Merck. This peptide was covered by Integra's patent.[27]

Due to Dr. Cheresh's breakthrough achievement, Merck and Scripps entered into a new collaboration agreement in September 1995 to fund the "necessary experiments to satisfy the biological bases and regulatory (FDA) requirements for the implementation of clinical trials" with EMD 66203 or a derivative thereof.[28] Dr. Cheresh then proceeded to conduct *in vivo* and *in vitro* experiments on EMD 66203 and two derivatives of it, EMD 85189 and EMD 121974, in order to evaluate each peptide as potential drug candidates. These "tests measured the efficacy, specificity, and toxicity of the particular peptides as angiogenesis inhibitors, and evaluated their mechanism of action and pharmacokinetics in animals."[29] Based on these tests, in November 1996 Merck's pharmaceutical steering committee selected EMD 85189 for pre-clinical development; in April 1997, Merck switched to EMD 121974 as its most promising candidate for clinical testing.[30] In October 1998, Merck reached an agreement with the National Cancer Institute (NCI) to sponsor the clinical trials, and later that year, the NCI filed an IND application with the FDA for EMD 121974.[31]

When Integra became aware of Merck's agreement with Scripps to conduct angiogenesis research for commercial purposes, Integra offered Merck the opportunity to purchase licenses to use its patented RGD peptides. In July 1996, after Merck had declined the offer, Integra sued Merck, Scripps, and Dr. Cheresh, seeking monetary damages for Merck's alleged patent infringement and a declaratory judgment against Scripps and Dr. Cheresh.[32] In defense, Merck asserted that its actions involving the RGD peptides came within the common-law research exemption[33] and the statutory safe harbor afforded by § 271(e)(1).

The District Court's Decision in *Integra*

At the conclusion of trial, the U.S. District Court for the Southern District of California dismissed Integra's claim for declaratory judgment and held that the common-law research exemption protected Merck's pre-1995 use of the RGD peptides.[34] However, the court found that a question of fact remained as to whether Merck's post-1995 actions fell within the scope of the § 271(e)(1) safe harbor. The district court instructed the jury that, for Merck to prevail on the safe harbor defense, it must prove by a preponderance of the evidence that it was objectively reasonable for the company to believe that "there was a decent prospect" that the experiments "would contribute, relatively directly," to the generation of information likely to be relevant to the drug approval regulatory process.[35]

The jury found Merck liable for infringing Integra's patents and that Merck had failed to show that § 271(e)(1) protected its post-1995 research activities. The jury awarded damages of $15 million in royalties. In response to post-trial motions, the district court dismissed Integra's suit against Scripps and Dr. Cheresh, but affirmed the jury's monetary award, explaining that there was substantial evidence to show that the connection between the experiments and FDA review was "insufficiently direct to qualify for the [§ 271(e)(1) exemption]."[36]

Integra in the Federal Circuit

In June 2003, a divided panel of the Court of Appeals for the Federal Circuit affirmed the district court's determination as to liability but reversed the court's refusal to modify the damages award.[37] The panel majority found that safe harbor does not "reach any exploratory research that may rationally form only a predicate for future FDA clinical tests."[38] In confining the § 271(e)(1) exemption to research activities that contribute "relatively directly" to information "reasonably related" to clinical testing for the FDA, the appellate court stated:

> In this case, the Scripps work sponsored by Merck was not clinical testing to supply information to the FDA, but only general biomedical research to identify new pharmaceutical compounds. The FDA has no interest in the hunt for drugs that may or may not later undergo clinical testing for FDA approval.[39]

Furthermore, the court expressed concern that construing the safe harbor provision more expansively "would effectively vitiate the exclusive rights of patentees owning biotechnology tool patents," since patented research tools are often used in general research to identify candidate drugs and experiments on those drugs.[40]

On January 7, 2005, the U.S. Supreme Court granted *certiorari* to review the court of appeals' interpretation of the safe harbor provision.[41]

THE U.S. SUPREME COURT'S DECISION IN *INTEGRA*

The question presented to the Supreme Court was "whether uses of patented inventions in preclinical research, the results of which are not ultimately included in a submission to the Food and Drug Administration (FDA), are exempted from infringement by 35 U.S.C. § 271(e)(1)."[42] In a unanimous opinion written by Justice Scalia, the Court vacated the judgment of the Federal Circuit and held that the § 271(e)(1) safe harbor protected the preclinical use of patented compounds "as long as there is a reasonable basis for believing that the experiments will produce 'the types of information that are relevant to an IND or NDA'" submission to the FDA.[43]

The Court explained:

[W]e think it apparent from the statutory text that § 271(e)(1)'s exemption from infringement extends to all uses of patented inventions that are reasonably related to the development and submission of *any* information under the FDCA. ... This necessarily includes preclinical studies of patented compounds that are appropriate for submission to the FDA in the regulatory process. There is simply no room in the statute for excluding certain information from the exemption on the basis of the phase of research in which it is developed or the particular submission in which it could be included.[44]

The Court rejected Integra's argument that the scope of the safe harbor is limited only to preclinical studies pertaining to the safety of a drug in humans.[45] Since the FDA requires an IND to be filed *before* human trials can begin, IND applications must include summaries of a drug's efficacy, pharmacokinetics, pharmacology, and toxicological effects in animals.[46] This data would necessarily have to be developed in preclinical studies — information that is "reasonably related" to an FDA submission and thus covered by § 271(e)(1).[47]

The Court further disagreed with Integra's claim that Merck's preclinical research is disqualified from safe harbor protection because the experiments were not conducted in conformity with the FDA's "good laboratory practices" (GLP) regulations. Two reasons supported the Court's reasoning: first, the FDA's GLP regulations concerning preclinical studies apply only to experiments on drugs "to determine their safety," and not to studies of a drug's efficacy, mechanism of action, pharmacology, or pharmacokinetics; second, even non-GLP compliant safety-related studies are suitable for submission in an IND, when such studies are accompanied by a reason for the noncompliance.[48]

Basic Research Not Protected

The Court placed an outer limit to the safe harbor provision by endorsing the Federal Circuit's conclusion that the exemption does not reach all experimental activity that at some point, however attenuated, may lead to an FDA approval process.[49] For example, safe harbor does not embrace basic scientific research performed on a patented compound without the intent to develop a particular drug or without a reasonable belief that the compound will cause a particular physiological effect that the researcher desires.[50] Thus, the boundary line between unprotected basic research and protected preclinical research is reached when a scientist discovers that a patented compound produces a "particular" physiological effect through a "particular" biological process.[51]

The Standard for "Reasonable Relation"

In denying safe harbor protection for Merck's preclinical activities, the Federal Circuit had relied upon the fact that the "Scripps-Merck experiments did not supply information for submission to the [FDA], but instead identified the best drug candidate to subject to future clinical testing under the FDA processes."[52] The Supreme Court dismissed the appellate court's narrow interpretation of the "reasonably related" requirement in § 271(e)(1). Such a construction, the Court explained, "disregards the reality that ... scientific testing is a process of trial and error," and that "neither the drugmaker nor its scientists have any way of knowing whether an initially promising candidate will prove successful over a battery of experiments."[53] Thus, under certain conditions, the Court noted that the safe harbor provision is "sufficiently broad" to protect the use of patented compounds in experiments that are not ultimately submitted to the FDA or drug experiments that are not ultimately the subject of an FDA submission.[54]

The Court announced a standard for construing § 271(e)(1)'s reasonable relation requirement in a way that "leaves adequate space for experimentation and failure on the road to regulatory approval":

> At least where a drugmaker has a reasonable basis for believing that a patented compound may work, through a particular biological process, to produce a particular physiological effect, and uses the compound in research that, if successful, would be appropriate to include in a submission to the FDA, that use is "reasonably related" to the "development and submission of information under ... Federal law."[55]

An Unresolved Question: Patented Research Tools

Research tools are defined as "tools that scientists use in the laboratory, including cell lines, monoclonal antibodies, reagents, animal models, growth factors, combinatorial chemistry and DNA libraries, clones and cloning tools (such as PCR), methods, laboratory equipment and machines."[56] Smaller biotechnology companies and universities that invent research tools are concerned that a broader construction of § 271(e)(1) encompassing these tools will deprive them of licensing fees that they collect from larger pharmaceutical firms.[57] Moreover, some companies rely on such fees for their financial existence, since many of these research tools have little commercial value beyond usage in drug research.[58] The Federal Circuit in *Integra* had specifically identified this potential negative consequence for patented research tools, in its support for a more limited safe harbor:

> [T]he context of this safe harbor originally keyed its use to facilitating expedited approval of patented pioneer drugs already on the market. Extending § 271(e)(1) to embrace all aspects of new drug development activities would ignore its language and context with respect to the [Hatch-Waxman Act] in an attempt to exonerate infringing uses only potentially related to information for FDA approval. Moreover, such an extension would not confine the scope of § 271(e)(1) to *de minimis* encroachment on the rights of the patentee. For example, expansion of § 271(e)(1) to include the Scripps-Merck activities would effectively vitiate the exclusive rights of patentees owning biotechnology tool patents. Thus, exaggerating § 271(e)(1) out of context would swallow the whole benefit of the Patent Act for some categories of biotechnological inventions. Needless to say, the [Hatch-Waxman Act] was [not] meant ... to deprive entire categories of inventions of patent protection.[59]

In its amicus curiae brief submitted to the Supreme Court, the U.S. Government suggests that § 271(e)(1) does not apply to patented research tools.[60] The Government's brief explains that the safe harbor section, by its own terms, applies only to "a patented invention." The Patent Act defines the term "invention" to mean any "invention or discovery," "*unless* the context otherwise indicates."[61] The brief asserts that the context of § 271(e)(1) indicates that Congress may not have intended to include patented research tools within the scope of the safe harbor exemption. Since most research tools are used to study or develop other compounds for submission to the FDA regulatory approval process, rather than being themselves the subject of FDA regulatory review, it is plausible to conclude that research tools are not "patented inventions" within the meaning of the statute.[62]

In *Integra,* the Supreme Court expressly declined to decide whether or to what extent the exemption applies to patented research tools since the matter was not at issue in the case. The Court explained that Integra had never argued that the RGD peptides were used by Merck/Scripps as research tools, "and it is apparent from the record that they were not."[63] Thus, without a definitive judicial determination from the Court, the use of patented research tools in drug research and development may or may not fall under the § 271(e)(1) exemption from infringement. Such uncertainty over the patent rights of makers of research tools could serve as a source of continued confusion and litigation in this area.

CONCLUDING OBSERVATIONS

The original legislative intent behind the Hatch-Waxman Act that created § 271(e)(1) was to facilitate the introduction of a generic drug upon the patent expiration of the brand name drug. However, as the Supreme Court explained in the *Eli Lilly* case that broadened § 271(e)(1) beyond generic drugs to the entire statutory scheme of FDA regulation:"[I]t is not the law that a statute can have no effects which are not explicitly mentioned in its legislative history."[64]

The consequences of the Supreme Court's decision in *Integra* are significant. Some observers argue that if the Federal Circuit's opinion had not been vacated, its narrow interpretation of the patent law's safe harbor potentially would have created a chilling effect on the development of innovative, pioneer drugs and new generic drugs. Limiting § 271(e)(1) to only clinical research appears contrary to the objectives of the Hatch-Waxman Act: If a drug manufacturer could not perform the preclinical studies needed to obtain FDA approval to conduct clinical studies, "the [§ 271(e)(1)] exemption would never be reached because the underlying preliminary research and development work could not be undertaken" without risking patent infringement liability.[65]

The Supreme Court's more expansive construction of § 271(e)(1) avoids this result. Since "it will not always be clear to parties setting out to seek FDA approval for their new product exactly which kinds of information, and in what quantities, it will take to win that agency's approval," the safe harbor provision is needed to immunize certain preclinical studies that use patented compounds.[66] The Court also provided an articulated standard for courts, scientists, drug companies, and patent holders to follow concerning the scope of § 271(e)(1) coverage: Safe harbor applies if there is a reasonable basis to believe that the preclinical experiments will produce information that is relevant to an IND or NDA submission with the FDA. Failure to meet this standard would constitute infringing conduct not exempted by § 271(e)(1). By unanimous opinion, the *Integra* Court has emphatically clarified that preclinical use of patented compounds in pharmaceutical research is not categorically unprotected and can qualify for the patent law's safe harbor as long as it comes within this enunciated standard.

However, the *Integra* Court left unresolved the issue of whether research tools come within the scope of the safe harbor exemption. It is important to note that *Integra* does not affect the validity and value of patented research tools when they are employed in basic research or for purposes unrelated to an FDA submission.[67] Yet the unauthorized use of research tools in the development of information for the FDA regulatory process may constitute infringing conduct or could be exempted by the patent law's safe harbor. This legal uncertainty raises concerns about the enforceability of research tool patents in this circumstance. Unless or until the Supreme Court answers this question in a future case, Congress may desire to clarify § 271(e)(1)'s applicability to research tools.

REFERENCES

[1] Integra Lifesciences I, Ltd. v. Merck KGaA, 331 F.3d 860, 867 (Fed. Cir. 2003) (citation omitted).

[2] Merck KGaA v. Integra Lifesciences I, Ltd., __ U.S. __, 125 S. Ct. 2372, 2383 (2005).

[3] *Id.* at 2380.

[4] 35 U.S.C. § 271(a).

[5] 35 U.S.C. § 271(e)(1).

[6] P.L. 98-417, 98 Stat. 1585 (1984), codified in 15 U.S.C. §§ 8b-68c, 70b; 21 U.S.C. §§ 301, 355, 360cc; 28 U.S.C. § 2201; and 35 U.S.C. §§ 156, 271, 282.

[7] The statutory exemption is also called the "Bolar Amendment " or "FDA exemption," since it effectively overturns the decision of the U.S. Court of Appeals for the Federal Circuit in Roche Products, Inc. v. Bolar Pharmaceutical Co., Inc., 733 F.2d 858 (Fed. Cir. 1984), which had found Bolar, a manufacturer of generic drugs, liable for infringing Roche's patented drug during the last six months of the term of the patent in its testing and investigation activities related to FDA drug approval requirements.

[8] *Roche,* 733 F.2d at 860.

[9] For more information regarding the Hatch-Waxman Act, see CRS Report RL30756, Patent Law and Its Application to the Pharmaceutical Industry: An Examination of the Drug Price Competition and Patent Term Restoration Act of 1984 ("The Hatch-Waxman Act"), by Wendy H. Schacht and John R. Thomas; and CRS Report RL32377, The Hatch-Waxman Act: Legislative Changes Affecting Pharmaceutical Patents, by Wendy H. Schacht and John R. Thomas.

[10] P.L. 75-717 (1938), codified in 21 U.S.C. §§ 301 et seq.

[11] 21 U.S.C. § 355(i). Generic drug companies may file an abbreviated new drug application (ANDA) with the FDA. 21 U.S.C. § 355(j). An ANDA must reveal that the generic product has the same active ingredients as, and is bioequivalent to, a prior approved brand name drug. Also, in its ANDA, the generic drug manufacturer may rely upon the safety and efficacy data of the original drug manufacturer.

[12] 21 U.S.C. § 355(i)(1)(A).

[13] 21 U.S.C. § 355(b)(1).

[14] For more information concerning the FDA drug approval process, *see* CRS Report RL32797, *Drug Safety and Effectiveness: Issues and Action Options After FDA*

[15] H.Rept. 98-857 (II) at 30, 98th Cong., 2d Sess. (1984), *reprinted in* 1984 U.S.C.C.A.N. 2686, 2714.
[16] 496 U.S. 661 (1990).
[17] *Id.* at 665.
[18] *Id.* at 667.
[19] *Id.* at 666.
[20] *Id.* at 674.
[21] James N. Czaban and Nishita Doshi, Supreme Court Applies Broad Interpretation of Bolar Amendment to Protect Innovative Drug Research From Claims of Patent Infringement, 70 PAT., TRADEMARK, and COPYRIGHT J. (BNA) 1726 (June 24, 2005).
[22] *Integra*, 331 F.3d at 862-63.
[23] *Id.* at 873 (Newman, J., dissenting).
[24] *Id.* at 863.
[25] *See* Merck is Not the Same as Merck, *available at* [http://www.merck.de/servlet/PB/menu/1014710/index.html].
[26] Telios Pharms., Inc. v. Merck KGaA, 1997 U.S. Dist. LEXIS 24187, Case No. 96-CV-1307 (S.D. Cal., Sept. 9, 1997), at *3.
[27] *Integra*, 125 S. Ct. at 2378 n.3.
[28] *Integra*, 331 F.3d at 863.
[29] *Integra*, 125 S. Ct. at 2378. Efficacy means how well a drug can be expected to work in curing a disease; mechanism of action is how it achieves those results; pharmacokinetics is the rate at which a drug is absorbed into and eliminated from the bloodstream; and toxicity is the negative side effects of the drug at different dosages. Brief for Petitioner Merck KGaA at 12-13, *Merck KGaA v. Integra Lifesciences I, Ltd.*, 125 S. Ct. 2372 (2005) (No. 03-1237).
[30] *Telios Pharms.*, 1997 U.S. Dist. LEXIS 24187, at *5.
[31] *Integra*, 331 F.3d at 873 (Newman, J., dissenting).
[32] *Id.* at 863.
[33] The common-law research exemption is a limited, judge-made exception to the patentee's right to exclude. Its historic foundations arise from Whittemore v. Cutter, 29 Fed. Cas. 1120, 1121 (C.C.D. Mass. 1813), in which Justice Story stated, "it could never have been the intention of the legislature to punish a man, who constructed such a machine merely for philosophical experiments..." Courts have recognized the use of this exemption for research that has no commercial purpose. *Integra*, 331 F.3d at 874-75 (Newman, J., dissenting).
[34] *Integra*, 125 S. Ct. at 2379.
[35] Id.
[36] *Id.* at 2380.
[37] On remand, the District Court reduced the award to $6.375 million, on the calculated basis of $1.5 million per year as a reasonable royalty between the infringement period August 1994 and November 1998. *Integra*, 2004 WL 2284001, at *11 (S.D. Cal. Sept. 7, 2004).
[38] *Integra*, 331 F.3d at 867.
[39] *Id.* at 866-67.

[40] *Id.* at 868.
[41] *Integra,* 125 S. Ct. 823 (2005).
[42] *Integra,* 125 S. Ct. at 2376.
[43] *Id.* at 2383-84 (citation omitted).
[44] *Id.* at 2380 (emphasis in original) (citations omitted).
[45] *Id.* at 2381 ("We do not understand the FDA's interest in information gathered in preclinical studies to be so constrained.")
[46] 21 C.F.R. § 312.23(a)(5).
[47] *Integra,* 125 S. Ct. at 2381.
[48] *Id.* at 2382 (citations omitted).
[49] Id.
[50] Id.
[51] *Id.* at 2383.
[52] *Integra,* 331 F.3d at 865-66.
[53] *Integra,* 125 S. Ct. at 2383.
[54] *Id.* at 2382. The legislative history of § 271(e)(1) supports this reasoning: "A party which develops [information for the FDA regulatory process], but decides not to submit an application for approval, is protected [by the safe harbor] as long as the development was done to determine whether or not an application for approval would be sought." H.Rept. 98-857 (I) at 45, 98th Cong., 2d Sess. (1984), *reprinted in* 1984 U.S.C.C.A.N. 2647, 2678.
[55] *Integra,* 125 S. Ct. at 2383 (citing § 271(e)(1)).
[56] Integra, 331 F.3d at 874 n.4 (citing Sharing Biomedical Research Resources: Principles and Guidelines for Recipients of NIH Research Grants and Contracts, 64 Fed. Reg. 72090, 72092 n.1 (Dec. 23, 1999)).
[57] David Savage and Denise Gellene, High Court Boosts Drug Research; Justices Say Companies Are Free to Use Patented Compounds in Developing Medicines. Analysts Say Ruling May Hurt Some Biotech Firms, L.A. TIMES, June 14, 2005, at C1.
[58] Id.
[59] *Integra,* 331 F.3d at 867.
[60] Brief for United States as *Amicus Curiae*, at 29, *Merck KGaA v. Integra Lifesciences I, Ltd.,* 125 S. Ct. 2372 (2005) (No. 03-1237).
[61] 35 U.S.C. § 100(a) (emphasis added).
[62] Brief for United States as *Amicus Curiae*, at 29.
[63] *Integra,* 125 S. Ct. at 2382 n.7.
[64] *Eli Lilly,* 496 U.S. at 669 n.2 (quoting Pittston Coal Group v. Sebben, 488 U.S. 105, 115 (1988)).
[65] Brief for United States as *Amicus Curiae*, at 14 (quoting Bristol-Myers Squibb Co. v. Rhone-Poulenc Rorer, Inc., 2001 WL 1512597, at *6 (S.D.N.Y. Nov. 28, 2001)).
[66] *Integra,* 125 S. Ct. at 2383 (citation omitted).
[67] Brief for United States as *Amicus Curiae*, at 30.

Chapter 5

FILE-SHARING SOFTWARE AND COPYRIGHT INFRINGEMENT: *METRO-GOLDWYN-MAYER STUDIOS, INC. V. GROKSTER, LTD.*[*]

Brian Yeh and Robin Jeweler

ABSTRACT

Metro-Goldwyn-Mayer Studios, Inc. v. Grokster, Ltd. is a Ninth Circuit Court of Appeals decision considering allegations of contributory and vicarious copyright infringement by companies which distribute peer-to-peer file-sharing software. The software facilitates direct copyright infringement by its users. It is the first decision to reject infringement claims against and find in favor of companies distributing the software. To date, other digital media file-sharing software decisions have found in favor of the copyright holders, most notably *A and M Records, Inc. v. Napster, Inc.* and *In re: Aimster Copyright Litigation*. But in *Grokster*, the court granted summary judgment for the software companies. This report provides a general overview of peer-to-peer file-sharing technology and then examines why *Grokster* produced a result different from those in other peer-to-peer software litigation. On December 10, 2004, the U.S. Supreme Court granted a writ of certiorari to hear an appeal in the *Grokster* case.

One explanation for *Grokster* is the differences in the technological design of the various peer-to-peer systems. While the pioneering file-sharing network Napster provided exclusively for the exchange of audio files, the software companies sued in *Grokster* employ more advanced peer-to-peer technology that allows the additional sharing of video clips, text documents, and computer programs. The *Grokster* court acknowledged these expanded capabilities as legitimate uses of the software, and thus became the first court to accept the "substantial, noninfringing uses" defense to copyright infringement liability, a defense developed by the U.S. Supreme Court in *Sony Corp. of America v. Universal City Studios, Inc.* Two months after the U.S. district court decision in *Grokster*, the Seventh Circuit Court of Appeals in *Aimster* also expressed qualified support for application of the *Sony* defense to file-sharing software, but nevertheless upheld a preliminary injunction against Aimster because the software company failed to

---

[*] Excerpted from CRS Report RL31998, dated December 17, 2004.

demonstrate that its peer-to-peer service was ever actually used for any substantial noninfringing purposes.

Another factor determinative to the *Grokster* court was the software companies' limited contribution to the infringing activity of its software users, and their limited ability to police their networks. Whereas Napster actively provided on-going services and technical support to its users in locating and downloading music files, the *Grokster* defendants distribute software that operates across peer-to-peer networks outside of their control and supervision. This more sophisticated software allows users to connect to each other and swap files directly, without the need for a centralized search index or website to facilitate the file transfers, as Napster had maintained.

BACKGROUND

File-sharing software programs that create "peer-to-peer" (P2P) network connections between computers enable the transmission of data and communications over the Internet. A variety of P2P programs, such as those offered by Grokster, Ltd., StreamCast Networks, Inc., and Kazaa BV, are typically available for free download from the distributors' websites. After installing a P2P program (called a "client application") onto the computer, the user runs the application to connect to the computers of other users of that particular P2P software who are currently "on-line." The client application allows users to "share" files located on their computer hard-drives. Once users make files available for sharing with each other, anyone who uses the same company's software to connect to the respective P2P network may locate and download desired files easily and at no cost. For example, a user of the Grokster software can directly access files saved on another Grokster user's computer hard-drive. Or a user can search for a particular file name, such as an MP3 song title, across all users' computers connected to the Grokster network, and then download a copy of that file onto his or her computer.

The motion picture and music recording industries brought several legal actions alleging copyright infringement against companies that distribute file-sharing software. Prior to *Grokster*, nearly all of the cases found in their favor, including *A and M Records, Inc. v. Napster, Inc.*[1] and *In re: Aimster Copyright Litigation.*[2] However, in *Metro-Goldwyn-Mayer Studios, Inc. (MGM) v. Grokster, Ltd.*[3][3] the defendant software distributors were found not liable for copyright infringement. *Grokster* is the first P2P file-sharing case to date involving allegations of music piracy where a court has shielded P2P software companies from liability.

MGM STUDIOS, INC. V. GROKSTER, LTD. IN U.S. DISTRICT COURT

Plaintiffs, twenty-eight organizations in the motion picture and music recording industries, sued several P2P companies for contributory and vicarious copyright infringement. The plaintiffs and defendant companies Grokster and StreamCast Networks filed cross-motions for summary judgment because "the only question before the [c]ourt (as to liability) is a legal one: whether [d]efendants' materially undisputed conduct gives rise to copyright liability."[4]

The court granted a summary judgment determining that defendant companies, Grokster and StreamCast, were not secondarily liable for copyright infringement committed by users of their P2P software (called Grokster and Morpheus, respectively).[5] Finding no evidence that the defendants had any material involvement in their users' infringing conduct, the court held that Grokster and StreamCast were not liable for contributory infringement.[6] Summary judgment was also warranted on the vicarious infringement claim, the court reasoned, because there was no evidence showing that the P2P companies had the right or ability to supervise the conduct of end-users of their products.[7]

Contributory Infringement

To succeed on a claim of contributory or vicarious infringement, plaintiffs must first demonstrate that at least some users of defendants' software are themselves engaged in *direct* infringement of plaintiff's copyrighted works. The district court declared that "many of those who use [d]efendants' software do so to download copyrighted media files, including those owned by [p]laintiffs, and thereby infringe [p]laintiffs' rights of reproduction and distribution."[8]

The concept of contributory infringement has its roots in tort law and the notion that one should be held accountable for directly contributing to another's infringement.[9] For contributory infringement liability to exist, a court must find that the secondary infringer "with knowledge of the infringing activity, induces, causes or materially contributes to the infringing conduct of another."[10]

Although defendants were "generally aware" that their software programs were being used by their customers for infringing activities, the court found that level of knowledge insufficient to establish liability for contributory infringement.[11] The court relied on the 1984 U.S. Supreme Court case, *Sony Corp. of America v. Universal City Studios, Inc.*,[12] which ruled that Sony's sale of video cassette recorders (VCRs) did not subject the manufacturer to liability for contributory copyright infringement. The Court explained that since VCRs are capable of both infringing and noninfringing uses, generic knowledge of and potential infringement were insufficient to impose liability on the Betamax VCR manufacturer, Sony.[13] The *Grokster* court analogized P2P software to the VCR, finding the former to be capable of substantial current and potential future noninfringing uses, such as searching for files available in the public domain or downloading content that has been authorized for distribution by the copyright holder.[14]

The court agreed with the software companies' argument that in order to be liable for contributory infringement, the requisite level of knowledge is actual knowledge of *specific* infringement at a time when either defendant materially contributes to the infringement, and can use that knowledge to stop it.[15] Although the plaintiffs had sent the defendants thousands of notices regarding infringing conduct, the notices are "irrelevant if they arrive when [d]efendants do nothing to facilitate, and cannot do anything to stop, the alleged infringement."[16]

The court also determined that neither Grokster nor StreamCast had satisfied the second element of contributory copyright infringement liability, namely, encouraging, assisting, or materially contributing to the infringing activity. Aside from distributing the P2P software, Grokster and StreamCast do not do anything actively to facilitate their users' infringing activity.[17] Using the Grokster and Morpheus client applications to connect to each other directly, individuals can send and receive files without having any network traffic pass through servers owned or controlled by defendant companies. This decentralized file-sharing network is distinctly different from the Napster system, in which all search requests went through and relied upon Napster's central servers. Applying the reasoning of *Fonovisa, Inc. v. Cherry Auction, Inc.*,[18] a case which found contributory copyright infringement by operators of a swap meet where vendors were selling counterfeit records to swap meet attendees, the *Napster* court determined that Napster materially contributed to the infringing conduct by providing the "site and facilities" for direct infringement. In contrast, the *Grokster* court stated that neither Grokster nor StreamCast provides the "site and facilities" for direct infringement, and the companies did not actively and substantially participate in the exchange of files between their users.[19]

Vicarious Infringement

Liability for vicarious copyright infringement is warranted in cases where the defendant "has a right and ability to supervise the infringing activity and also has a direct financial interest in such activities."[20] The district court noted that although the defendants offered their P2P software free to the public, they derive substantial revenue from advertising. The more people who download the software, the higher the revenue generated. The defendants therefore had a "direct financial interest" in the infringing conduct.[21]

Nonetheless, in the court's view, Grokster and StreamCast lacked the ability to supervise or control their users' conduct because they could not terminate users or restrict access to the P2P networks. The court distinguished the defendants' software from the Napster integrated "system." While Napster possessed the ability to monitor and police its file-sharing network, and often exercised its ability to exclude particular users from the network, the *Grokster* defendants merely distribute the P2P software and have no similar power over the end-users of the software once it has passed into their hands.[22]

GROKSTER IN THE NINTH CIRCUIT COURT OF APPEALS

At the outset, the court noted that the question of direct infringement, although undisputed, was not before it. Rather, the issue was one of secondary, i.e., contributory and/or vicarious copyright infringement liability, and it fully concurred in the district court's "well reasoned analysis" that defendants were not liable under either theory.

Contributory Infringement

The three elements required to prove contributory copyright infringement are (1) direct infringement by a primary infringer, (2) knowledge of the infringement, and (3) material contribution to it. The Court of Appeals generally tracked the reasoning of the lower court and concluded that the architecture of the P2P systems precluded the defendants from having requisite knowledge of and making a material contribution to copyright infringement.

The court invoked the "staple article of commerce" doctrine from patent law, cited by the Supreme Court in *Sony,* for the proposition that "it would be sufficient to defeat a claim of contributory copyright infringement if the defendant showed that the product was 'capable of substantial' or 'commercially significant noninfringing uses.'"[23] With respect to the "knowledge"element, as interpreted by the Ninth Circuit in its *Napster* decision, this means:

> [I]f a defendant could show that its product was capable of substantial or commercially significant noninfringing uses, then constructive knowledge of the infringement could not be imputed. Rather, if substantial noninfringing use was shown, the copyright owner would be required to show that the defendant had reasonable knowledge of specific infringing files.[24]

Because the court found that there is no question that P2P software is capable of substantial noninfringing uses, it would *not* impute constructive knowledge of infringement by others to the defendants. They must have "reasonable knowledge of specific infringement" to satisfy the knowledge requirement. And, the copyright owners needed to establish that the software distributors had specific knowledge of the infringement at a time in which they contributed to it and failed to act upon their knowledge. This knowledge standard was not met by the plaintiffs' after-the-fact notices of infringement. By the time they arrived, the software distributors could not prevent it.

In the court's view, the software distributors did not, indeed, could not materially contribute to infringing activity as a consequence of the P2P program structure, discussed below. The distributors do not provide a "site and facility" for the infringement, the infringing material does not reside on the defendants' computers, nor do they have the ability to suspend user accounts.

Vicarious Infringement

In order to establish vicarious infringement, a copyright owner must demonstrate (1) direct infringement by a primary party, (2) a direct financial benefit to the defendant, and (3) the right and ability to supervise the infringers.[25] Concurring with the district court, the Court of Appeals found it to be undisputed that there was both direct infringement and the defendants benefitted financially from the software via advertising revenue. But "the right and ability to supervise infringers" was missing. The court reiterated the lower court's finding that the communication between the defendants and their users does not provide a point of access for filtering or searching for infringing files since infringing material and index information do not pass through the defendants' computers.

The Court of Appeals considered and rejected the argument that "a right to supervise" could be interpreted to mean that the distributors had a responsibility to alter the software to prevent users from sharing copyrighted files. Only where one has already been determined to be liable for vicarious infringement does a duty to exercise "police" powers arise. It also rejected what it characterized as a "blind eye" theory of liability for vicarious infringement. Copyright owners argued that "turning a blind eye to detectable acts of infringement for the sake of profit" should give rise to liability.[26] But the court found that there is no separate "blind eye" theory or element of vicarious liability that exists independently of the traditional elements of liability.

In closing, the Court of Appeals echoed the district court in declining to interpret copyright liability law to achieve public policy goals that might protect copyright owners' economic interests but alter general copyright law in profound ways with unknown ultimate consequences. That function is the prerogative of the Congress.

GROKSTER COMPARED TO NAPSTER AND AIMSTER

In the *Napster* and *Aimster* cases, the courts found that the plaintiff record companies would likely prevail on their contributory and vicarious infringement claims and granted preliminary injunctions against the defendant companies.[27] The distinction in the outcome in *Grokster* is based, in large part, on the technological differences in the design of the file-sharing networks in question and, to some extent, judicial uncertainty over the reach of the standards articulated by the U.S. Supreme Court in *Sony*.

P2P Network Architecture

Unlike Napster and Aimster, Grokster and Morpheus software users connect to file-sharing networks with no central database server. Instead, Grokster provides for dynamic "root supernodes," which are a group of randomly chosen computers which are connected to the P2P network at a particular time. The Grokster software "self-selects" its supernode status for the day; a user's computer can function as a supernode one day and not on the following day.[28] As a supernode, a connected P2P user's computer acts as an index server, collecting information about shared files located on other users' computers. "Normal" Grokster clients connect to their "neighborhood" supernode to perform searches for files.[29] This creates a "two-tiered" organizational distribution structure for the P2P network, with groups of regular Grokster users clustered around a single supernode.[30] All search traffic and information passes through these personal computers acting as supernodes, none of which are owned or controlled by Grokster.

StreamCast's Morpheus network operates in an even more decentralized fashion. While Grokster licenses proprietary FastTrack networking technology from Sharman Networks, StreamCast bases its Morpheus program on a non-proprietary, "open-source"[31] technology called Gnutella.[32] The Gnutella P2P network can be accessed using not only Morpheus software, but also other Gnutella-based software distributed by companies such as "BearShare" and "LimeWire." Unlike Grokster's supernode architecture, search requests on the Gnutella network quickly pass from user to user until a matching file is found or until the search request expires. The query can reach over 8,000 other computers on the Gnutella network before expiring.[33]

While Aimster and Napster actively assisted their end-users in locating specific files and facilitating the download transactions, Grokster did not. Aimster provided an illustrated tutorial on its website which "methodically demonstrated how to transfer and copy copyrighted works over the Aimster system."[34] The Court of Appeals in *Aimster* referred to this tutorial as "an invitation to infringement," overtly encouraging Aimster users to infringe copyrights.[35] Aimster also offered its users a $4.95 monthly subscription service called "Club Aimster," which presented a list of the "top forty" most frequently downloaded songs by Aimster users. This list included a "play" button next to each song that allowed users to click on it to download the music file from wherever it was located in the Aimster network to the requesting member's computer.[36]

Napster similarly provided substantial assistance to its users through its server, maintained as a collective directory of shared MP3 files available on each connected user's computer. Any Napster user could run a search on this centralized index for a desired song title in order to transfer a copy of the file from the "host user" (the individual sharing the file) to the requesting user's hard-drive.[37]

Therefore, when the Napster network was terminated, its demise prevented any further P2P file-sharing among its users. But, because Grokster and StreamCast operate decentralized networks, "if either [company] closed their doors and deactivated all computers within their control, users of their products could continue sharing files with little or no interruption."[38] According to the *Grokster* court, this is a "seminal distinction" between Grokster/StreamCast and Napster which formed the basis for its judgment that the defendants did not materially contribute to infringing activity.[39]

SUBSTANTIAL, NONINFRINGING USE

In *Sony*, the U.S. Supreme Court found that despite the fact that VCRs could be used by consumers to infringe copyrights, manufacturers were not liable for contributory copyright infringement because the VCR was capable of "substantial noninfringing uses."[40] The Court identified private, noncommercial "time-shifting" of broadcast television as a noninfringing fair use of the technology.[41] It emphasized that the primary use of the VCR was fair, and, in addition, it had the potential to be used for time-shifting sports, educational, religious and other programming that was authorized for copying or was copied without objection from the copyright holder.[42]

As a result of *Sony*, "substantial, noninfringing uses" became a judicially recognized defense to claims of contributory infringement. The defense arises from the Court's dicta that "[t]he sale of copying equipment, like the sale of other articles of commerce, does not constitute contributory infringement if the product is widely used for legitimate, unobjectionable purposes... [I]t need merely be capable of substantial noninfringing uses."[43] The Seventh and Ninth Circuits disagree, however, on the scope and proper application of *Sony*'s legal principles.

The *Grokster* court determined that the defendants' P2P software satisfied the *Sony* test for substantial noninfringing uses. Analogizing P2P software companies to manufacturers of VCRs and copy machines, the court noted that the Grokster and Morpheus programs are used for both lawful and unlawful ends, and therefore the software distributors cannot be liable without evidence of active and substantial contribution to the infringement itself.[44] Compared to the Napster file-sharing network, which *functioned primarily* to exchange copyrighted audio files, the Morpheus and Grokster software *may be used* to transfer a variety of different file types.[45] The more versatile capability of the *Grokster* defendants' P2P software was a critical factor in the courts' decisions. But the Court of Appeals also takes a somewhat rigid and formulaic approach to the application of *Sony's* mandates.

In comparison, the Seventh Circuit in *Aimster* expressed a more limited and flexible view of *Sony's* scope. It found that the *Sony* doctrine is applicable to software services as well as to "articles of commerce," but was inclined to weigh and balance the factors necessary to establish contributory infringement. For example, while it agreed with the recording industry "that the ability of a service provider to prevent its customers from infringing *is a factor to be considered* in determining whether the provider is a contributory infringer...[i]t is not necessarily a controlling factor[.]"[46] Unlike the Ninth Circuit, which categorically rejected the idea of "willful blindness" as a factor in determining whether an infringer has actual knowledge" of infringement, the Seventh Circuit ruled that "[w]illful blindness is knowledge, in copyright law ... as it is in the law generally."[47]

And, with respect to "substantial noninfringing use," the Court of Appeals in *Aimster* rejected the music industry's argument that the *Sony* defense could not withstand anything more than a "mere showing that a product may be used for infringing purposes." In fact, it went so far as to say that actual knowledge of specific infringing uses alone could constitute a sufficient condition for deeming a facilitator liable for contributory infringement.[48] Nor would it accept that *Sony* confers immunity from contributory infringement liability when software is capable of substantial noninfringing use:

> We ... do not buy Aimster's argument that ... all Aimster has to show in order to escape liability for contributory infringement is that its file-sharing system *could* be used in noninfringing ways, which obviously it could be. Were that the law, the seller of a product or service used *solely* to facilitate copyright infringement, though it was capable in principle of noninfringing uses, would be immune from liability for contributory infringement. That would be an extreme result, and one not envisaged by the *Sony* majority.[49]

Thus, although a product or service may be physically capable of noninfringing uses, the Seventh Circuit Court of Appeals emphasized the degree of probability that the software is or would be employed for noninfringing use. Specifically, "[w]hen a supplier is offering a product or service that has noninfringing as well as infringing uses, some estimate of the respective magnitudes of these uses is necessary for a finding of contributory infringement."[50]

Grokster takes a broader reading of the "capability" aspect of *Sony*. It rejects the "primary use" analysis applied by the *Aimster* court, and focuses on the fact that Grokster and Morpheus are being used, and *can be used*, for substantial noninfringing purposes.

CONCLUSION

Since the district court's decision in *Grokster* cast doubt on the viability of legal action against P2P software companies, copyright holders focused their enforcement efforts against direct infringers themselves. Despite the "impracticability or futility of a copyright owner's suing a multitude of individual infringers,"[51] the music recording industry is currently pursuing exactly that course of action, filing lawsuits against particularly egregious P2P swappers of copyrighted music.[52] Whether this effort by the music industry will significantly slow illegal file-sharing or help to deter future infringement remains to be seen. Another unknown is the reaction of consumers to these legal attacks aimed directly at individual P2P software users.

The *Aimster* and *Grokster* decisions indicate that copyright owners seeking to enforce their rights in P2P file-swapping litigation may encounter more difficulty overcoming legal obstacles to proving contributory copyright infringement. Since Napster, the next generation of P2P technologies has become increasingly decentralized, permitting file exchange between computer users without significant intermediary assistance by a software company. This P2P software also is "dual-use" in character – capable and increasingly utilized for both noninfringing and infringing purposes – raising the possibility of greater success by software companies in invoking the *Sony* defense to claims of contributory copyright infringement.

Largely as a consequence of *Napster*, some assert that the new generation of P2P companies deliberately designed their software to exploit perceived "loopholes" in the application of traditional copyright law to emergent digital technologies. Indeed, the district court in *Grokster* acknowledged this possibility, noting that the "[d]efendants may have intentionally structured their businesses to avoid secondary liability for copyright infringement, while benefitting financially from the illicit draw of their wares."[53]

The ultimate impact of *Grokster* on copyright law in uncertain. Some Members of Congress reacted to the *Grokster* decision by introduction of S. 2560, the Inducing Infringement of Copyrights Act of 2004, 108th Cong., 2d Sess. (2004). This bill would codify criteria to establish secondary liability for copyright infringement. Perhaps as a consequence of the divergent interpretations of *Sony* by the Ninth and Seventh Circuits, the U.S. Supreme Court has agreed to review the *Grokster* decision. In doing so, it may clarify the relationship between *Sony* and secondary liability for copyright infringement.

REFERENCES

[1] 239 F.3d 1004 (9th Cir. 2001).
[2] 252 F.Supp.2d 634, 648 (N.D. Ill. 2002), *aff'd,* 334 F.3d 643 (7th Cir. 2003).
[3] 259 F.Supp.2d 1029 (C.D. Ca 2003) *aff'd,* 380 F.3d 1154 (9th Cir.), *cert. granted,*___ S.Ct. ___, 2004 WL 2289054, 73 U.S.L.W. 3247 (U.S. Dec 10, 2004) (No. 04-480).
[4] 259 F.Supp.2d at 1031.
[5] This summary judgment only applies to software offered by defendants Grokster and StreamCast Networks. Another defendant, Sharman Networks, distributor of Kazaa Media Desktop software, was not a party to these summary judgment motions. *See id.* at 1033.
[6] *Id.* at 1043.
[7] *Id.* at 1046.
[8] *Id.* at 1034-1035.
[9] Fonovisa, Inc. v. Cherry Auction, Inc., 76 F.3d 259, 264 (9th Cir. 1996).
[10] *Napster,* 239 F.3d at 1019.
[11] *Grokster,* 259 F.Supp.2d at 1037-1038.
[12] 464 U.S. 417 (1984).
[13] *Grokster,* 259 F.Supp.2d at 1035, citing *Sony,* 464 U.S. at 442.

[14] The court cited, for example, "distributing movie trailers, free songs, or other non-copyrighted works; using the software in countries where it is legal; or sharing the works of Shakespeare." *See id.*
[15] *Id.* at 1038.
[16] *Id.* at 1037.
[17] *Id.* at 1039-1043.
[18] *Fonovisa, supra* note 9. The court found that swap meet organizers materially contributed to the infringing activity of the vendors by providing "support services" such as booth space, parking, utilities, advertising, plumbing, and customers.
[19] *Grokster,* 259 F.Supp.2d at 1041-1043.
[20] *Fonovisa,* 76 F.3d at 262, citing Gershwin Publishing Corp. v. Columbia Artists Management, Inc., 443 F2d. 1159, 1162 (2d. Cir. 1971).
[21] "Just as customers were attracted to the swap meet in *Fonovisa*, because of the sale of counterfeit goods, individuals are attracted to Defendants' software because of the ability to acquire copyrighted material free of charge." *Grokster,* 259 F.Supp.2d at 1044.
[22] *Id.* at 1045.
[23] *Grokster,* 380 F.3d at 1160.
[24] *Id.* (Citations omitted.)
[25] *Id.* at 1164.
[26] *Id.* at 1166.
[27] The Aimster service was renamed "Madster" in January 2002, after a ruling by the National Arbitration Forum panel that the Internet domain name "aimster.com" violated the trademark for America Online (AOL)'s instant messaging service. To avoid confusion, this report will continue to refer to the file-sharing service as "Aimster." *See* [http://www.usatoday.com/tech/news/2002/02/01/aimster-now-madster.htm].
[28] *Grokster,* 259 F.Supp.2d at 1040.
[29] [http://www.grokster.com/helpfaq.html#Is%20Grokster%20a%20distributed%20network]. Only connected computers that are particularly powerful or have fast Internet connections are chosen as "supernodes."
[30] *Grokster,* 259 F.Supp.2d at 1040.
[31] Open-source refers to any software program whose "source code" (the software programming language) is made freely available to the public for use or modification by other software developers. *See* [http://www.opensource.org/docs/definition.php].
[32] *Grokster,* 259 F.Supp.2d at 1041.
[33] *See* [http://computer.howstuffworks.com/file-sharing.htm/printable].
[34] *Aimster,* 252 F.Supp.2d at 643.
[35] *Aimster,* 334 F.3d at 651.
[36] *Id.* at 652.
[37] *Napster,* 239 F.3d at 1012.
[38] *Grokster,* 259 F.Supp.2d at 1041.
[39] Id.
[40] *Sony,* 464 U.S. at 456.
[41] Time-shifting refers to the practice of recording a broadcast television program to view it at a later time, and then erasing it afterwards. *See Sony,* 464 U.S. at 423.
[42] *Id.* at 444.

[43] *Sony,* 464 U.S. at 442.
[44] *Grokster,* 259 F.Supp.2d at 1043.
[45] StreamCast produced evidence that its Morpheus program "is regularly used to facilitate and search for public domain materials, government documents, media content for which distribution is authorized, media content as to which the rights owners do not object to distribution, and computer software for which distribution is permitted." *Id.* at 1035.
[46] *Aimster,* 334 F.3d at 648. (Emphasis supplied.)
[47] *Id.* at 650.
[48] *Id.* at 649.
[49] *Id.* at 651. (Emphasis in original.)
[50] *Id.* at 649.
[51] *Id.* at 645.
[52] *See generally,* [http://www.riaa.com/news/newsletter/062503.asp].
[53] *Grokster,* 259 F.Supp.2d at 1046.

Chapter 6

COOPERATIVE R&D: FEDERAL EFFORTS TO PROMOTE INDUSTRIAL COMPETITIVENESS[*]

Wendy H. Schacht

ABSTRACT

In response to the foreign challenge in the global marketplace, the United States Congress has explored ways to stimulate technological advancement in the private sector. The government has supported various efforts to promote cooperative research and development activities among industry, universities, and the federal R&D establishment designed to increase the competitiveness of American industry and to encourage the generation of new products, processes, and services. Among the issues before Congress are whether joint ventures contribute to industrial competitiveness and what role, if any, the government has in facilitating such arrangements.

Collaborative ventures are intended to accommodate the strengths and responsibilities of all sectors involved in innovation and technology development. Academia, industry, and government often have complementary functions. Joint projects allow for the sharing of costs, risks, facilities, and expertise.

Cooperative activity covers various institutional and legal arrangements including industry-industry, industry-university, and industry-government efforts. Proponents of joint ventures argue that they permit work to be done that is too expensive for one company to support and allow for R&D that crosses traditional boundaries of expertise and experience. Such arrangements make use of existing, and support the development of new, resources, facilities, knowledge, and skills. Opponents argue that these endeavors dampen competition necessary for innovation. Federal efforts to encourage cooperative activities include the National Cooperative Research Act; the National Cooperative Production Act; tax changes permitting credits for industry payments to universities for R&D and deductions for contributions of equipment used in academic research; and amendments to the patent laws vesting title to inventions made under federal funding in universities. Technology transfer from the government to the private sector is facilitated by several laws. In addition, there are various ongoing cooperative programs supported by various federal departments and agencies.

---

[*] Excerpted from CRS Report IB89056, dated June 3, 2005.

Given the increased popularity of cooperative programs, questions might be raised as to whether they are meeting expectations. It may be too soon to determine the effectiveness of the joint R&D venture as a mechanism to increase technological advancement in the United States. There is often a long time lag between research and the availability of a product, process, or service. Many of the collaborative activities fostered by the federal government are of recent origin and therefore have not had sufficient time to generate measurable results. However, raising certain issues might serve to develop a framework for addressing future, near-term decisions concerning technology development and cooperative R&D. These include questions about the emphasis on collaborative ventures in research rather than in technology development; cooperative manufacturing; defense vs. civilian support; and access by foreign companies.

MOST RECENT DEVELOPMENTS

Over the past 25 years, congressional initiatives have promoted cooperative research and development among industry, universities, and the federal R&D establishment. This is evident in legislation creating technology transfer mechanisms as well as in support for the Advanced Technology Program (ATP) and the Manufacturing Extension Partnership (MEP) at the National Institute of Standards and Technology (NIST). However, many of these activities were revisited beginning in the 104th Congress given the Republican majority's statements in favor of indirect measures such as tax policies, intellectual property rights, and antitrust laws to promote technology development; increased government support for basic research; and decreased direct federal funding for private sector technology initiatives. Although none of the cooperative programs have been terminated, several were funded at reduced levels. Since FY2000, the original appropriation bills passed by the House contained no support for ATP and in FY2004 MEP financing decreased 63% from the previous fiscal year. However, the FY2005 Omnibus Appropriations Act (P.L. 108-447) restored funding to MEP to fund existing centers. The legislation financed MEP at $107.5 million and ATP at $136.5 million (after a mandated 0.8% across the board rescission and a 0.54% rescission from the Commerce, Justice, State discretionary accounts). Also passed by the 108th Congress, the CREATE Act, P.L. 108-453, made changes in patent law to promote cooperative research among industry, government, and academia. P.L. 108-311 extended the research tax credit through December 31, 2005. In the 109th Congress, H.R. 250, introduced January 6, 2005 and reported to the House, amended, on May 23, 2003, establishes several new manufacturing technology programs for small and medium-sized firms. S. 296 would authorize appropriations for MEP through FY2008, including $115 million for FY2006. The Administration's FY2006 budget request proposes $46.8 million for the Manufacturing Extension Partnership and no funding for the Advanced Technology Program. Several bills, including H.R. 1454 and H.R. 1736, would make the research and experimentation tax credit permanent. S. 1020 would extend the credit through the end of 2007.

BACKGROUND AND ANALYSIS

Rationale

In response to concerns over competition from foreign firms, the U.S. Congress has increasingly looked for ways the federal government can stimulate technological innovation in the private sector. This technological advancement is critical in that it contributes to economic growth and long term increases in our standard of living. New technologies can create new industries and new jobs; expand the types and geographic distribution of services; and reduce production costs by making more efficient use of resources. The development and application of technology also plays a major role in determining patterns of international trade by affecting the comparative advantages of industrial sectors. Since technological progress is not necessarily determined by economic conditions, it can have effects on trade independent of shifts in macroeconomic factors that may affect the marketplace.

Joint ventures are an attempt to facilitate technological advancement within the industrial community. Academia, industry, and government can play complementary roles in technology development. While opponents argue that cooperative ventures stifle competition, proponents assert that they are designed to accommodate the strengths and responsibilities of these sectors. Collaborative projects attempt to utilize and integrate what the participants do best and to direct these efforts toward the goal of generating new goods, processes, and services for the marketplace. They allow for shared costs, shared risks, shared facilities, and shared expertise.

The lexicon of current cooperative activity covers various different institutional and legal arrangements. These ventures might include industry-industry joint projects involving the creation of a new entity to undertake research, the reassignment of researchers to a new effort, and/or hiring new personnel. Collaborative industry-university efforts may revolve around activities in which industry supports centers (sometimes cross-disciplinary) for research at universities, funds individual research projects, and/or exchanges personnel. Cooperative activities with the federal government might include projects that use federal facilities and researchers, federal funding for industry-industry or industry-university efforts, or financial support for centers of excellence at universities to which the private sector has access.

There are many different types of cooperative arrangements. The flexibility associated with this concept can allow for the development of institutional and organizational plans tailored to the specific needs of the particular project. Issues of patent ownership, disclosure of information, licensing, and antitrust are to be resolved on an individual basis within the general guidelines established by law governing joint ventures.

Collaborative ventures can be structured either "horizontally" or "vertically." The former involves efforts in which companies work together to perform research and then use the results of this research within their individual organizations. The latter involves activities where researchers, producers, and users work together. Both approaches are seen as ways to address some of the perceived obstacles to the competitiveness of American firms in the marketplace.

Joint Industrial Research

Traditionally, the federal government has funded research and development to meet mission requirements; in areas where the government is the primary user of the results; and/or where there is an identified need for R&D not being performed in the private sector. Most government support is for basic research which is often long-term and highly risky for individual companies; yet research can be the foundation for breakthrough achievements which can revolutionize the marketplace. Studies have shown that inventions based on R&D are the more important ones. However, the societal benefits of research tend to be greater than those that can be captured by the firm performing the work. Thus the rationale for federal funding of research in industry.

The major emphasis of legislative activity has been on augmenting research in the industrial community. This focus is reflected in efforts to encourage companies to undertake cooperative research arrangements and expand the opportunities available for increases in research activities. Collaboration permits work to be done which is too expensive for one company to fund and also allows for R&D that crosses traditional boundaries of expertise and experience. A joint venture makes use of existing, and supports development of new resources, facilities, knowledge, and skills.

The concentration on increased research as a prelude to increased technological advancement was based upon the "pipeline model" of innovation. This process was understood to be a series of distinct steps from an idea through product development, engineering, testing, and commercialization to a marketable product, process, or service. Thus increases at the beginning of the pipeline — in research — were expected to result in analogous increases in innovation at the end. However, this model is no longer considered valid. Innovation is rarely a linear process and new technologies and techniques often occur that do not require basic or applied research or development. Most innovations are actually incremental improvements to existing products and processes. In some areas, particularly biotechnology, research is closer to a commercial product than this conception would indicate. In others, the differentiation between basic and applied research is artificial. The critical factor is the commercialization of the technology. Economic benefits accrue only when a technology or technique is brought to the marketplace where it can be sold to generate income and/or applied to increase productivity.

In the recent past, it was increasingly common to find that foreign companies were commercializing the results of U.S. funded research at a faster pace than American firms. In the rapidly changing technological environment, the speed at which a product, process, or service is brought to the marketplace is often a crucial factor in its competitiveness. The recognition that more than research needs to be done has lead to other approaches at cooperative efforts aimed at expediting the commercialization of the results of the American R&D endeavor. These include industry-university joint activities, use of the federal laboratory system by industry, and industry-industry development efforts where manufacturers, suppliers, and users work together.

Industry-University Cooperative Efforts

Industry-university cooperation in R&D is one important mechanism intended to facilitate technological innovation. Traditionally, universities perform much of the basic research integral to certain technological advancements. They are generally able to undertake fundamental research because it is part of the educational process and because they do not have to produce for the marketplace. The risks attached to work in this setting are fewer than those in industry where companies must earn profits. Universities also educate and train the scientists, engineers, and managers employed by companies.

Academic institutions do not have the commercialization capacity available in industry and necessary to translate the results of research into products and processes that can be sold in the marketplace. Thus, if the work performed in the academic environment is to be integrated into goods and services, a mechanism to link the two sectors must be available. Prior to World War II, industry was the primary source of funding for basic research in universities. This financial support helped shape priorities and build relationships. However, after the war the federal government supplanted industry as the major financial contributor and became the principal determinant of the type and direction of the research performed in academic institutions. This situation resulted in a disconnection between the university and industrial communities. Because industry and not the government is responsible for commercialization, the difficulties in moving an idea from the research stage to a marketable product or process appear to have been compounded.

Efforts to encourage increased collaboration between the academic and industrial sectors might be expected to augment the contribution of both parties to technological advancement. Company support for research within the university provides additional funds and information on the concerns and direction of industry. For many companies, access to expertise and facilities outside of the firm expands or complements available internal resources. Yet, such cooperation should not necessarily be seen as a panacea. Oftentimes, collaborative ventures fail because of various factors including conflicting goals, differing research cultures, and financial disagreements.

Federal Laboratory-Industry Interaction

The federal government can share its extensive facilities, expertise, knowledge, and new technologies with partners in a cooperative venture. In certain cases, the government laboratories have scientists and engineers with experience and skills, as well as equipment, not available elsewhere. The government also has a vested interest in technology development. It does not have the mandate or resources to manufacture goods but has a stake in the availability of products and processes to meet mission requirements. In addition, technological advancement contributes to the economic growth vital to the health and security of the nation.

Collaboration between government laboratories and industry is not, however, just a one way street. In several technological areas, particularly electronics and computer software, the private sector is more advanced in technologies important to the national defense and welfare of this country. Interaction with industry offers federal scientists and engineers valuable information to be used within the government R&D enterprise.

FEDERAL INITIATIVES IN COOPERATIVE R&D

The cooperative venture concept is not new. In the early 1970s, the National Science Foundation established its Industry-University Cooperative Research Centers program. The Electric Power Research Institute, a research organization supported by the electric power utilities, has been in operation since 1973. In the private sector, the Microelectronics and Computer Technology Corporation (MCC), which performs research for its member firms, and the Semiconductor Research Corporation (SRC), which funds research in universities, were created in the early 1980s. The difference today is the number of projects and the scope of legislative activity designed to promote cooperative ventures.

Faced with pressures from foreign competition, the government's interest appears to be expanding beyond that of funding R&D, to meeting other critical national needs including the economic growth that flows from new commercialization in the private sector. While acknowledging that the commercialization of technology is the responsibility of the business community, in the past several years the government has attempted to stimulate innovation and technological advancement in industry. These activities often involve the removal of barriers to technology development in the private sector, thereby permitting market forces to operate and the provision of incentives to encourage increased innovation related efforts in industry. Cooperative R&D efforts are a part of both these trends.

The National Cooperative Research Act (P.L. 98-462) is designed to encourage companies to undertake joint research which is typically long-term, risky, and often too expensive for one company to finance. This legislation clarifies the antitrust laws and requires that the "rule of reason" standard be applied in determinations of violations of these laws; that cooperative research ventures are not to be judged illegal "per se". It also eliminates treble damage awards for those research ventures found in violation of the antitrust laws if prior disclosure (as defined in the law) has been made. In addition, P.L. 98-462 makes some changes in the way attorney fee awards are made to discourage frivolous litigation against joint research ventures without simultaneously discouraging suits of plaintiffs with valid claims. Over 750 joint ventures have filed with the Justice Department since this law was enacted.

P.L. 103-42, the National Cooperative Production Amendments Act of 1993, amends the National Cooperative Research Act by, among other things, extending the original law's provisions to joint manufacturing ventures. These provisions are only applicable, however, to cooperative production when the principal manufacturing facilities are "located in the United States or its territories, and each person who controls any party to such venture...is a United States person, or a foreign person from a country whose law accords antitrust treatment no less favorable to United States persons than to such country's domestic persons with respect to participation in joint ventures for production."

The Omnibus Trade and Competitiveness Act of 1988 (P.L. 100-418) created the Advanced Technology Program (ATP) at the Department of Commerce's National Institute of Standards and Technology. This program provides seed funding, matched by private sector investment, to companies or consortia comprised of universities, companies, and/or government laboratories for the development of generic technologies that have broad application across industrial sectors. As of May 2004, 736 projects have been funded representing approximately $2.2 billion in federal financing matched by $2 billion in financing from the private sector. Of these 736 projects, 211 were or are joint ventures. Eleven initial R&D programs were selected for funding, almost half of which involved consortia. Twenty-seven awards were made to programs in the second year; approximately one-third were *consortia*. In December 1992, 21 new ATP awards were made, including three joint ventures. Thirty additional projects were funded in 1993, and, in October 1994, 41 awards were made in four key technology areas: information infrastructure for healthcare; tools for DNA diagnostics; component-based software; and computer-integrated manufacturing for electronics. Fourteen are cooperative efforts. In November 1994, 47 additional awards were made in the general competition and in the area of manufacturing composite structures. Twenty-four involve collaborative R&D. Of the 24 awards announced on July 13, 1995, 35% of the projects in the general competition were joint ventures and 29% in the focused competition. The following month 21 additional awards were made of which 9 were cooperative efforts. In early September, another 44 grants were awarded including 19 joint ventures. Later in that month, 10 more awards were made of which three were to cooperative efforts. On January 25, 1996, an additional four projects received awards; three involved multiple firms. In March 1997, NIST announced that it would fund 8 new proposals from the FY1996 general competition of which 2 were collaborative projects. Sixty-four awards were made in October 1997; 15 involving multiple companies. In October 1998, NIST awarded funding for 79 new projects involving more than 150 companies, 11 universities, and several federal laboratories. This reflects changes in the ATP selection criteria designed to encourage large companies to participate in joint ventures with small firms and academic institutions. Thirty-seven awards for FY1999 were made on October 7, 1999. Of these, 27 are either joint ventures or involve additional organizations working as subcontractors. In FY2001, 13 of the 59 grants involved collaborative projects while in FY2002, 10 of the 61 awards went to joint ventures. Of the 16 awards made in July 2003, 3 were for collaborative projects. In September of 2003, 44 awards were made of which 9 were joint ventures. An additional 27 awards were made in May 2004, four involving cooperative activities. (For more information, see CRS Report 95-36 SPR, *The Advanced Technology Program*.)

Appropriations for the Advanced Technology Program were $35.9 million in FY1991, $47.9 million in FY1992, and $67.9 million in FY1993. FY1994 appropriations expanded significantly to $199.5 million and even further to $431 million in FY1995. However, P.L. 104-6, the DOD Emergency Supplemental Appropriations and Rescissions Act, rescinded $90 million of this amount. The President's FY1996 budget request for ATP was $490.9 million. There were no authorizations. The original appropriations bill, H.R. 2076, which passed the Congress but was vetoed by the President, provided no financing for ATP. The final appropriations legislation, P.L. 104-134, funded the Advanced Technology Program at $221 million for FY1996. The following year, FY1997, the Omnibus Consolidated Appropriations Act (P.L. 104-208) provided support levels of $225 million, but $7 million was rescinded by P.L. 105-18. While no authorization legislation was enacted for FY1998, P.L. 105-119 funded ATP at $192.5 million. The President's FY1999 budget request included $259.9 million for this program, an increase of 35%. However, P.L. 105-277, the Omnibus Consolidated Appropriations Act, funded ATP at $197.5 million, 3% above the previous year. This figure reflected a $6 million rescission to account for "deobligated" funds resulting from prior projects that had been terminated early.

In the FY2000 budget, the Administration requested $238.7 million for ATP, an increase of 21% over FY1999. No authorizations were enacted. S. 1217, as passed by the Senate on

July 22, 1999, would have appropriated $226.5 million, 15% more than the previous year. H.R. 2670, as passed by the House on August 5, 1999, contained no appropriated funding for ATP. The report accompanying the House bill stated that "...the program has not produced a body of evidence to overcome those fundamental questions about whether the program should exist in the first place." Yet, P.L. 106-113, the Consolidated Appropriations Act, provided the Advanced Technology Program with $142.6 million for FY2000, financing that was 28% below the level of the previous year. For FY2001, the Clinton Administration requested ATP funding of $175.5 million, an increase of 23% over prior year funding. The original appropriations bill, as passed by the House, again provided no support for the program. P.L. 106-553, does fund the ATP at $145.7 million for FY2001, 2% above the previous fiscal year.

The Bush Administration's FY2002 budget proposed suspending all funding for new ATP awards pending an evaluation of the program. However, $13 million would have been provided to meet financial commitments for on-going projects. H.R. 2500, as first passed by the House, provided no support for new ATP projects but did include $13 million to fund prior year commitments. The original Senate-passed version of H.R. 2500 would have funded the program at $204.2 million. P.L. 107-77 finances ATP at $184.5 million, a 27% increase over FY2001.

In the FY2003 budget, the President requested $108 million for the Advanced Technology Program. This figure was 35% below the FY2002 appropriation. A number of Continuing Resolutions supported the program at FY2002 levels until the 108th Congress passed P.L. 108-7 which appropriated $178.8 million in FY2003 (after the 0.65% across the board mandated by the legislation).

The Administration's FY2004 budget included $27 million for ATP to cover on-going commitments; no new projects would be funded. H.R. 2799, the appropriations bill first passed by the House on July 23, 2003, contained no funding for ATP. However, H.R. 2799 was subsequently incorporated into H.R. 2673, which became P.L. 108-199, the FY2004 Consolidated Appropriations Act. This legislation finances the program at $170.5 million (after a mandated rescission). As reported to the Senate from the Committee on Appropriations, S. 1585 would have provided $259.6 million for ATP.

For FY2005, the President's budget proposal, as well as H.R. 4754, the FY2005 appropriations bill originally passed by the House on July 8, 2004, contained no funding for ATP. As reported to the Senate by the Committee on Appropriations, S. 2809 would have financed the program at $203 million, an increase of 19% over the previous fiscal year. The FY2005 Omnibus Appropriations Act, P.L. 108-447, provides the Advanced Technology Program with $136.5 million (after several rescissions mandated in the legislation), a decrease of 20% from FY2004.

The President's FY2006 budget request does not include support for ATP.

Several laws have attempted to facilitate industry-university cooperation. Title II of the Economic Recovery Tax Act of 1981 (P.L. 97-34) provided, in part, a temporary 25% tax credit for 65% of all company payments to universities for the performance of basic research. Firms were also permitted a larger tax deduction for charitable contributions of equipment used in scientific research at academic institutions. The Tax Reform Act of 1986 (P.L. 99-514) kept this latter provision, but reduced the credit for university basic research to 20% of all corporate expenditures for this work over the sum of a fixed research floor plus any decrease in non-research giving.

The 1981 Act also provided an increased charitable deduction for donations of new equipment by a manufacturer to an institution of higher education. This equipment must be used for research or training for physical or biological sciences within the United States. The tax deduction was equal to the manufacturer's cost plus one-half the difference between the manufacturer's cost and the market value, as long as it does not exceed twice the cost basis.

This research and experimentation tax credit expired in June 1992 when an extension contained in H.R. 11, the Enterprise Zone Tax Act, was vetoed by former President Bush. The Omnibus Budget Reconciliation Act, P.L. 103-66, reinstated the credit through July 1995 and made it retroactive to the date of its previous expiration. The credit again expired. However, P.L. 104-188, the Small Business Job Protection Act, reinstated the tax credit for application between July 1, 1996 and May 31, 1997. The Taxpayer Relief Act of 1997, P.L. 105-34, extended the credit for 13 months from June 1, 1997 through June 30, 1998. The tax credit expired once again but was reinstated through June 30, 1999, by P.L. 105-277. Several bills also were introduced that would have permitted the research tax credit to be applied to support for certain collaborative research consortia. The 106th Congress once again extended the credit. Title V of P.L. 106-170 reinstates the research and experimentation tax credit through June 30, 2004 and increases the credit rate applicable under the alternative incremental research credit by one percentage point per step. P.L. 108-311 extends the research credit through December 31, 2005.

Amendments to the patent and trademark laws contained in P.L. 96-517 also were designed to foster interaction between academia and the business community. This law provides, in part, for title to inventions made by contractors receiving federal R&D funds to be vested in the contractor if it is a university, not-for-profit institution, or a small business. Certain rights to the patent are reserved for the government and these organizations are required to commercialize within a predetermined and agreed upon time frame. Providing universities with patent title is expected to encourage licensing to industry where the technology can be manufactured or utilized, thereby creating a financial return to the academic institution. University patent applications and licensing have increased since this law was enacted. (For more discussion on this topic see CRS Report RL32076, *The Bayh-Dole Act: Selected Issues in Patent Policy and the Commercialization of Technology*, CRS Report RL30320, *Patent Ownership and Federal Research and Development: A Discussion on the Bayh-Dole Act and the Stevenson-Wydler Act* and CRS Report 98-862, *R&D Partnerships and Intellectual Property: Implications for U.S. Policy*.)

Many cooperative industry-industry or industry-university programs are supported and/or organized by the federal departments and agencies. These include, but are not limited to, the National Science Foundation's Engineering Research Centers, the approximately 40 Industry-University Cooperative Research Programs, and the more recent Science and Technology Centers. A program to match small businesses interested in joint manufacturing technology efforts has been created in the Department of Commerce.

While most legislative activities are intended to facilitate technological advance across industries, there have been several recent efforts to provide direct assistance for cooperative ventures in a particular industry. These initiatives are based, in part, on national defense and economic security concerns over specific technologies that are, or are perceived as, potentially critical to a wide range of businesses. Among the joint ventures, funded primarily by the Department of Defense, have been SEMATECH (a joint private sector semiconductor manufacturing research effort which is now privately financed), the National Center for

Manufacturing Sciences, and the steel initiative. In addition, DOD supports the Software Engineering Institute and the Department of Energy assists in the Partnership for a New Generation Vehicle initiative that, among other things, encourages joint R&D between federal laboratories and private firms leading to commercialization.

Cooperation between industry and the federal R&D enterprise is another facet of the effort to increase industrial competitiveness through joint ventures. The federal government will spend an estimated $83 billion for research and development in FY2000 to meet the mission requirements of the federal departments and agencies. This has led to many technologies and techniques, as well as to the generation of knowledge and skills, which may have applications beyond their original intent. To foster their development and commercialization in the industrial community, various laws have established institutions and mechanisms to facilitate the movement of ideas and technologies between the public and private sectors.

The Stevenson-Wydler Technology Innovation Act (P.L. 96-480), as amended by the Federal Technology Transfer Act (P.L. 99-502) and the Department of Defense FY1990 Authorizations (P.L. 101-189), provides, in part, a legislative mandate for technology transfer from the federal government to the private sector, establishes a series of offices in the agencies and/or laboratories to administer transfer efforts, provides incentives for federal laboratory personnel to actively engage in technology transfer, and creates new contractual means for industry to work with the laboratories including cooperative research and development agreements. P.L. 104-113, the National Technology Transfer and Advancement Act, attempts to clarify existing policy with respect to the dispensation of intellectual property under a CRADA by amending the Stevenson-Wydler Act. P.L. 106-404, the Technology Transfer Commercialization Act, makes changes in current practices concerning patents held by the government to make it easier for federal agencies to license such inventions to the private sector for commercialization. (For additional information see CRS Issue Brief IB85031, *Technology Transfer: Use of Federally Funded Research and Development*.)

The CREATE Act, P.L. 108-453, makes changes in the patent laws to promote cooperative research and development among universities, government, and the private sector. The bill amend section 103(c) of title 25, United States Code, such that certain actions between researchers under a joint research agreement will not preclude patentability.

The Omnibus Trade and Competitiveness Act (P.L. 100-418) established a program of regional Centers for the Transfer of Manufacturing Technology (now part of the Manufacturing Extension Partnership effort) to facilitate the movement to the private sector of knowledge and technologies developed under the aegis of the National Institute of Standards and Technology. (For more discussion, see CRS Report 97-104, *Manufacturing Extension Partnership Program*.) In addition, the law required that NIST provide technical assistance to state technology extension programs in an effort to improve private sector access to federal technology. (For additional Information, see CRS Issue Brief IB91132, *Industrial Competitiveness and Technological Advancement: Debate over Government Policy*.) Government-industry collaboration is further facilitated by a provision of the FY1991 National Defense Authorization Act (P.L. 101- 510) that amends Stevenson-Wydler to allow government agencies and laboratories to develop partnership intermediary programs to augment the transfer of laboratory technology to the small business community.

A pilot activity under the Small Business Development Act of 1992, the Small Business Technology Transfer program, facilitates cooperative work between small companies and federal labs leading to the commercialization of new technology. Scheduled to sunset in FY1996, the program was extended for one year until P.L. 105-135 reauthorized it through FY2001. Passed in the current Congress, P.L. 107-50 extends the STTR activity through FY2009 and increases the set-aside used to fund the program to 0.3% beginning in FY2004 when the amount of money available of individual Phase II grants increases to $750,000 (see CRS Report 96-402, *Small Business Innovation Research Program*).

ISSUES

It is not yet known whether federal support of cooperative ventures signals a long-term commitment to the development of technology. The former Clinton Administration set out a policy to actively promote joint R&D activities utilizing both direct and indirect federal support for expanded cooperative work leading to commercialization. However, given current concerns over the federal budget, it is unlikely that large sums of government money will be forthcoming for such efforts in the future. However, other actions may reflect federal interest in the process of technological advancement. The use of the extensive government R&D system, with its expensive state-of-the-art facilities, can provide both academia and industry with resources that may be beyond their financial ability. And despite the often short-term focus of budget decisions, federal funds and non-monetary contributions to cooperative ventures may be leveraged by contributions from state and local agencies and the private sector.

If the proliferation of programs is any indication, state and local jurisdictions have been in the forefront of cooperative endeavors. Many state and local economic development activities focus on increasing innovation and the use of technology in the private sector. Instead of competing for companies to relocate, many of these jurisdictions now see additional benefits accruing from the creation of new firms and the modernization of existing ones through the application of new technology. Various states and localities are attempting to foster an entrepreneurial climate by undertaking the development and support of a variety of programs to assist existing high technology businesses, to promote the establishment of new companies, and to facilitate the use of new technologies and processes in traditional industries. While these efforts vary by state and locality, many of them include industry-university-government cooperation. Several of the former President's proposals for increasing cooperative ventures built upon existing state and local activities in these areas. (For additional discussion, see CRS Report 96-958 SPR, *Technology Development: Federal-State Issues* (out of print; contact author for copies, 202-707-7066) and CRS Report 98-859, *State Technology Development Strategies: The Role of High Tech Clusters*.)

Proponents of cooperative work argue that certain benefits are associated with joint ventures. The increased popularity of this concept, and expanding federal support for this approach, however, might suggest some questions be raised to assess whether cooperative ventures are meeting expectations. Are there drawbacks to this effort in general and in specific instances? Are cooperative projects addressing the problems associated with the competitiveness of U.S. industry? Are they moving technology development in the right direction?

It might be expected that an increasing number of industries and/or companies will come to the federal government for assistance in supporting cooperative R&D activities. Despite opposition by some to what has been described as "picking winners or losers," various sectors of the government have chosen to provide funding for cooperative ventures in specific industries while requiring that the private sector generate matching funds. At the same time, there are programs and policies that attempt to facilitate cooperative efforts across industry in general. Decisions might need to be made whether one approach is better than the other, or if both should continue.

If part of government policy is to respond to individual industry requests for assistance, Congress may wish to consider developing procedures to select between industries and/or companies competing for limited federal funds. Can, and should, federal guidelines be established? In addition, is it possible to determine at this time what type of cooperative ventures are the most effective and efficient? Is there, in fact, one best model or should each venture be tailored to the specific situation? And finally, what are the implications of these decisions for policymaking in Congress?

Development

As noted above, innovation is a dynamic process that can involve idea origination, research, development, commercialization, and diffusion throughout the economy. However, it is not a linear process and an innovation may occur without developing through these steps. In fact, most innovations are actually incremental changes in existing goods and services in response to unmet market needs. The most crucial factor is the availability or use of the technology or technique in the marketplace.

In the recent past, the commercialization and diffusion of products and processes often stood out as significant problems in terms of the ability of U.S. industries to compete. Firms in several other countries, particularly Japan and the East Asian newly industrializing countries, have been successful in commercializing the results of R&D. In various instances, this was research initially performed in the United States, as evidenced by the VCR and semiconductor chips. Basic research and the pursuit of science are done successfully in the United States as indicated, in part, by the number of Nobel prizes awarded to Americans. However, excellence in science does not necessarily assure leadership in world markets. It has been noted that the United States was the world's premiere economic power in the 1920s when this nation was far from being in the forefront of science. Instead, market leadership is significantly affected by the development and application of technology to make the goods and services the consumers want to purchase.

Thus, questions may be raised as to whether programs and policies encouraging increased cooperative research, without concomitant efforts to facilitate the development and commercialization of technologies and techniques, can be effective mechanisms to increase the competitiveness of American industry. Do we need to know more about how to encourage the application of the research resulting from joint ventures in the manufacture of products and processes and in the delivery of services? Do these cooperative activities include mechanisms to facilitate the effective and timely transfer of the results back to the companies where they can be developed into goods for the marketplace? Since the major portion of the costs associated with bringing out a new product occur at the development and marketing stages, not in the research phase, should there be additional government incentives to encourage companies to spend funds for commercialization in addition to research?

Manufacturing

It is in the manufacturing arena where American companies appear to be the most vulnerable to foreign competition. Process technologies (those used in manufacturing) can significantly lower the costs of production and increase the quality of goods and services. In *Global Competition*, the President's Commission on Industrial Competitiveness (under former President Reagan) concluded that ". . . competitive success in many industries today is as much a matter of mastering the most advanced manufacturing processes as it is in pioneering new products."

The costs associated with the development and purchase of new manufacturing equipment are high. This is particularly true for the 350,000 small companies which make up a major segment of the manufacturing community. Several of the cooperative efforts supported by the federal government address these manufacturing concerns. The Manufacturing Technology program of the Department of Defense, the Advanced Manufacturing Technical Initiative of the Department of Energy, and the Manufacturing Extension Centers operated by the National Institute of Standards and Technology, although all different, are examples of government activities devoted to facilitating the development of new manufacturing techniques and their use in industry.

Considering the importance of manufacturing, the existing cooperative programs may not be sufficient to increase the competitiveness of American industry. Are there more effective types of joint ventures? Cooperative efforts, where resources could be pooled and the equipment shared, may be one way to improve the manufacturing capability of U.S. firms, large or small. Will joint manufacturing prove to be a viable option? Should existing cooperative manufacturing programs in certain agencies be expanded or should new efforts in other departments be developed? Should one government agency have the lead in policy determinations; if so, which federal department?

Defense vs. Civilian Support

Many of the industries interested in cooperative ventures with federal financial support have approached the Department of Defense and, to a lesser extent, the Department of Energy's Defense Programs because these agencies have the greatest amount of available resources and/or funding. They also tend to have the expertise to operate large-scale programs and maintain close ties with certain industrial sectors which could be encouraged to increase cooperation. In addition, both DOD and DOE have a vested interest in the availability of certain technologies which could be provided by a healthy domestic commercial market. However, questions remain whether sponsorship of certain cooperative ventures by DOD and the Department of Energy's defense-related programs will lead to increased commercialization in the civilian marketplace.

Critics argue that defense spending is not an effective mechanism to increase industry's ability to innovate and develop new technologies. Much of the research and development in the defense arena may be too specialized, overdesigned, and/or too costly to have value for commercial markets. The R&D also tends to concentrate on weapon systems and other defense hardware rather than on process technologies that are often necessary to improve manufacturing productivity. One reason cited for the competitive problems of the machine tool industry was its focus on defense needs rather than on the commercial market which is larger in the aggregate.

On the other hand, the U.S. commitment to military R&D has contributed to a favorable balance of trade in the defense and aerospace industries. In the SEMATECH effort, the purpose of DOD support was to facilitate the *commercial* development of technologies with critical defense applications. The companies involved in SEMATECH were experienced semiconductor manufacturers and were knowledgeable about the markets' needs and operations. Thus, although the initial work performed by this semiconductor consortium may have been partially funded by the Defense Advanced Research Project Agency, it was designed to result in new products and processes in the civilian marketplace where both defense and commercial demand can be met. SEMATECH now operates without direct federal financing.

The issue of cooperative work between the Defense Department and the private sector leading to commercial technologies was addressed in the former Technology Reinvestment Project and is part of the more recent Dual-Use Partnership Project. The Department of Energy has been expanding cooperative R&D activities in Defense Program laboratories in conjunction with an increase in all DOE collaborative efforts with industry. Recent significant decreases in the technology transfer budgets may impeded this effort, but several DOE defense laboratories are actively pursuing joint ventures with industry. (See CRS Report 98-81, *Cooperative Research and Development Agreements and Semiconductor Technology: Issues Involving the "DOE-Intel CRADA".*)

Access by Foreign Firms

With worldwide communications systems, it is virtually impossible to prevent the flow of scientific and technical information. What is critical to competitiveness is the speed at which this knowledge is used to make products, processes, and services for the marketplace. However, it appears that many foreign firms are willing and able to take the results of research performed both in the United States and their own countries and rapidly make high quality commercial goods. Many of these companies are purchasing American businesses or establishing U.S. subsidiaries to access American expertise. With the increased activity in research consortia, particularly those with federal support, questions might be asked as to whether or not foreign companies should or could be barred from access to the results. A larger issue is how to define an "American company." Is it determined by majority ownership, manufacturing, location, value added to the U.S. economy, or by some other definition? In addition, since technology is most effectively transferred by person-to-person interaction, would cooperative activities between American industry and foreign firms produce an outflow of information which could be used to increase competitive pressures?

Direct vs. Indirect Support

Government efforts to facilitate cooperative ventures have included both indirect supports and direct federal funding. Indirect measures include such things as tax policies, intellectual property rights, and antitrust laws that create incentives for the private sector. Other initiatives include government financing (on a cost shared basis) of joint efforts such as the Advanced Technology Program and Manufacturing Extension Partnerships. In the past, participants in the legislative process generally did not make definite (or exclusionary) choices between these two approaches. However, these activities were revisited in the 104th Congress given apparent Republican preferences for the funding of basic research and not technology development. For example, efforts to eliminate the Advanced Technology Program, funding for flat panel displays, and agricultural extension reflected concern over the role of government in developing commercial technologies and generally resulted in reductions of direct federal financing for such public-private partnerships. Issues were again raised in the subsequent Congresses although no relevant, on-going program was terminated. As the 109th Congress begins its budget deliberations, the future of cooperative R&D may be expected to be explored further. (For more information, see CRS Report 95-50, *The Federal Role in Technology Development*.)

108TH CONGRESS: LEGISLATION

P.L. 108-7, H.J.Res. 2

Omnibus FY2003 Appropriations Act. Among other things, funds the Advanced Technology Program at $180 million and the Manufacturing Extension Partnership at $106.6 million. Introduced January 7, 2003; referred to the House Committee on Appropriations. Passed House on January 8, 2003. Passed Senate, amended, on January 23, 2003. House and Senate agreed to conference report on February 13, 2003. Signed into law by the president on February 20, 2003.

P.L. 108-199, H.R. 2673

FY2004 Consolidated Appropriations Act. Among other things, funds the Advanced Technology Program at $170.5 million in FY2004 and the Manufacturing Extension Partnership at $38.7 million (with the 0.59% mandated rescission). H.R. 2673 reported to the House as an original measure from the House Committee on Appropriations July 9, 2003. Passed the House July 14, 2003 and passed the Senate November 6, 2003. House agreed to conference report on December 8, 2003. Senate agreed to conference report on January 22, 2004. Signed into law by the President on January 23, 2004.

P.L. 108-311, H.R. 1308

Amends the Internal Revenue Code of 1986 to extend the research tax credit through December 31, 2005, among other things. Introduced March 18, 2004; referred to the House Committee on Ways and Means. Passed House March 19, 2004 and passed the Senate with an amendment on June 5, 2004. Both the House and the Senate agreed to the conference report on September 23, 2004. Signed into law by the President on October 4, 2004.

P.L. 108-447, H.R. 4818

FY2005 Omnibus Appropriations Act. Among other things, funds ATP at $136.5 million in FY2005 and MEP at $107.5 million after several rescissions mandated in the legislation. Introduced July 13, 2004; referred to the House Committee on Appropriations. Passed the House on July 15, 2004. Passed the Senate, amended, on September 23, 2004. Conference report agreed to in both the House and Senate on November 20, 2004. Signed into law on December 8, 2004.

P.L. 108-453, S. 2192

The CREATE Act. Amends patent law to promote cooperative research among universities, government, and the private sector. Introduced March 10, 2004; referred to the Senate Committee on the Judiciary. Reported to the Senate on April 29, 2004. Passed Senate on June 25, 2004. Passed House on November 20, 2004. Signed into law on December 10, 2004.

LEGISLATION

H.R. 250 (Ehlers)

Manufacturing Technology Competitiveness Act. Creates an interagency committee to coordinate federal manufacturing R&D. Establishes and authorizes funding for a pilot collaborative manufacturing research grants program to promote the development of new manufacturing technologies through cooperative applied research among the private sector, academia, states, and other non-profit institutions. Mandates and authorizes financing for a manufacturing fellowship program. Creates and authorizes support for a manufacturing extension center competitive grants program to focus on new or emerging manufacturing technologies. Authorizes funding for the Manufacturing Extension Partnership, among other things. Introduced January 6, 2005; referred to the Committee on Science. Reported to the House, amended, May 23, 2005.

H.R. 1454 (Sensenbrenner)

Amends the Internal Revenue Code of 1986 to make the research and experimentation tax credit permanent. Introduced March 20, 2005; referred to the House Committee on Ways and Means.

H.R. 1736 (Johnson, N.)

Amends the Internal Revenue Code of 1986 to make the research and experimentation tax credit permanent, among other things. Introduced April 20, 2005; referred to the House Committee on Ways and Means.

S. 296 (Kohl)

Authorizes appropriations for the Manufacturing Extension Partnership through FY2008, including $115 million for FY2006. Introduced February 3, 2005; referred to the Senate Committee on Commerce, Science, and Transportation.

S. 1020 (Coleman)

"COMPETE" Act of 2005. Extends the research and experimentation tax credit through the end of 2007, among other things. Introduced May 12, 2005; referred to the Senate Committee on Finance.

In: Patent Technology
Editor: Juanita M. Branes, pp. 137-153

ISBN: 978-1-60021-220-8
© 2007 Nova Science Publishers, Inc.

Chapter 7

TECHNOLOGY TRANSFER: USE OF FEDERALLY FUNDED RESEARCH AND DEVELOPMENT[*]

Wendy H. Schacht

ABSTRACT

The government spends approximately one third of the $83 billion federal R&D budget for intramural research and development to meet mission requirements in over 700 government laboratories (including Federally Funded Research and Development Centers). The technology and expertise generated by this endeavor may have application beyond the immediate goals or intent of federally funded R&D. This can be achieved by technology transfer, a process by which technology developed in one organization, in one area, or for one purpose is applied in another organization, in another area, or for another purpose. It is a way for the results of the federal R&D enterprise to be used to meet other national needs, including the economic growth that flows from new commercialization in the private sector; the government's requirements for products and processes to operate effectively and efficiently; and the demand for increased goods and services at the state and local level.

Congress has established a system to facilitate the transfer of technology to the private sector and to state and local governments. Despite this, use of federal R&D results has remained restrained, although there has been a significant increase in private sector interest and activities over the past several years. Critics argue that working with the agencies and laboratories continues to be difficult and time-consuming. Proponents of the current effort assert that while the laboratories are open to interested parties, the industrial community is making little effort to use them. At the same time, State governments are increasingly involved in the process. At issue is whether incentives for technology transfer remain necessary, if additional legislative initiatives are needed to encourage increased technology transfer, or if the responsibility to use the available resources now rests with the private sector.

---

[*] Excerpted from CRS Report IB85031, dated July 1, 2005.

MOST RECENT DEVELOPMENTS

Past Administrations have made expanded use of federal laboratories and industry-government cooperation integral to efforts associated with technology development. In support of this approach, Congress enacted various laws facilitating cooperative research and development agreements (CRADAs) between federal agencies and the private sector, and increasing funding for technology transfer activities included in the Advanced Technology Program (ATP) and the Manufacturing Extension Partnership (MEP) at the National Institute of Standards and Technology (NIST), a laboratory of the Department of Commerce. However, many of these efforts have been revisited since the 104th Congresses, reflecting the Republican majority's preferences for indirect measures such as tax policies, intellectual property rights, and antitrust laws to promote technology development rather than to direct federal funding of private sector technology initiatives. While none of the relevant programs have been terminated, several have seen significant decreases in their budgets. Since FY2000. the original appropriation bills passed by the House did not contain any funding for ATP and in FY2004 support for MEP was cut by 63% from the previous fiscal year. During the last Congress, P.L. 108-199 provided FY2004 funding of $170.5 million for ATP and reduced financing for the Manufacturing Extension Partnership to $38.7 million (including the 0.59% rescission mandated in the legislation). The President's FY2005 budget request had no funding for ATP and would have financed MEP at $39.2 million. However, P.L. 108-447, the FY2005 Omnibus Appropriations Act, provided $136.5 million for ATP and $107.5 million for MEP (after a mandated 0.8% across the board rescission and a 0.54% rescission from the Commerce, Justice, State discretionary accounts). The CREATE Act, P.L. 108-453, made changes in patent law to promote cooperative research among universities, government, and the private sector. P.L. 108-311 extended the research tax credit through December 31, 2005. In the 109th Congress, H.R. 250, introduced January 6, 2005 and reported, amended, to the House on May 23, 2005, establishes several new manufacturing technology programs for small and medium-sized firms. S. 296, introduced February 3, 2005, would authorize appropriations for MEP through FY2008, including $115 million in FY2006. The Administration's FY2006 budget request proposes $46.8 million for MEP and no funding for ATP. H.R. 2862, the FY2006 State, Science, Commerce, Justice appropriations legislation, as passed by the House on June 16, 2005, would fund MEP at $106 million and offers no financing for ATP. The version of H.R. 2862 reported to the Senate from the Committee on Appropriations would also support MEP at $106 million but provides $140 million for ATP. Several bills, including H.R. 1454 and H.R. 1736, would make the research and experimentation tax credit permanent. S. 1020 would extend the credit through the end of 2007.

BACKGROUND AND ANALYSIS

The federal government spends approximately $83 billion per year on research and development to meet the mission requirements of the federal departments and agencies. Approximately one-third of this is spent for intramural research and development (R&D) by federal laboratories (including support for Federally Funded Research and Development

Centers). While the major portion of this activity has been in the defense arena, government R&D has led to new products and processes for the commercial marketplace including, but not limited to, antibiotics, plastics, airplanes, computers, microwaves, and bioengineered drugs. Given the increasing competitive pressures on U.S. firms in the international marketplace, proponents of technology transfer argue that there are many other technologies and techniques generated in the federal laboratory system which could have market value if further developed by the industrial community. Similarly, the knowledge base created by the agencies' R&D activities can serve as a foundation for additional commercially relevant efforts in the private sector.

The movement of technology from the federal laboratories to industry and to state and local governments is achieved through technology transfer. Technology transfer is a process by which technology developed in one organization, in one area, or for one purpose is applied in another organization, in another area, or for another purpose. In the defense arena it is often called "spin-off." Technology transfer can have different meanings in different situations. In some instances, it refers to the transfer of legal rights, such as the assignment of patent title to a contractor or the licensing of a government-owned patent to a private firm. In other cases, the transfer endeavor involves the informal movement of information, knowledge, and skills through person-to-person interaction. The crucial aspect in a successful transfer is the actual use of the product or process. Without this, the benefits from more efficient and effective provision of goods and services are not achieved. However, while the United States has perhaps the best basic research enterprise in the world — as evidenced in part by the large number of Nobel Prizes awarded to American scientists — other countries sometimes appear more adept at taking the results of this effort and making commercially viable products to be sold in U.S. and world markets. (For further discussion of innovation and economic growth, see CRS Issue Brief IB91132, *Industrial Competitiveness and Technological Advancement: Debate Over Government Policy.*)

Despite the potential offered by the resources of the federal laboratory system, the commercialization level of the results of federally funded research and development remained low through the 1980s. Studies indicated that only approximately 10% of federally owned patents were ever used. There were many reasons for this, including the fact that many of these technologies and patents have no commercial application. A major factor in successful transfer is a perceived market need for the technology or technique. However, because federal laboratory R&D is generally undertaken to meet an agency's mission or because there are insufficient incentives for private sector research that the government deems in the national interest, decisions reflect public sector, rather than commercial needs. Thus, transfer often depends on attempts to ascertain commercial applications of technologies developed for government use — "technology push" — rather than on "market pull." In other words, a technology is developed and a use for it established because the expertise exists rather than because it is needed.

Additional barriers to transfer involve costs. Studies have estimated that research accounts for approximately 25% of expenditures associated with bringing a new product or process to market. Thus, while it might be advantageous for companies to rely on government-funded research, there are still significant added costs of commercialization after the transfer of technology has occurred. However, industry unfamiliarity with these technologies, the "not invented here" syndrome, and ambiguities associated with obtaining title to or exclusive license for federally owned patents also contribute to a limited level of commercialization. Complicating the issue is the fact that the transfer of technology is a complex process that involves many stages and variables. Often the participants do not know or understand each other's work environment, procedures, terminology, rewards, and constraints. The transfer of technology appears to be most successful when it involves one-to-one interaction between committed individuals in the laboratory and in industry or state and local government. "Champions" are generally necessary to see a transfer through to completion because it is so often a time- and energy-consuming process. Given this, technology transfer is best approached on a case-by-case basis that can take into account the needs, operating methods, and constraints of the involved parties.

TECHNOLOGY TRANSFER TO PRIVATE SECTOR: FEDERAL INTEREST

The federal interest in the transfer of technology from government laboratories to the private sector is based on several factors. The government requires certain goods and services to operate. Much of the research it funds is directed at developing the knowledge and expertise necessary to formulate these products and processes. However, because the government has neither the mandate nor the capability to commercialize the results of the federal R&D effort, it must purchase technologies necessary to meet mission requirements from the private sector. Technology transfer is a mechanism to get federally generated technology and technical know-how to the business community where it can be developed, commercialized, and made available for use by the public sector.

Federal involvement in technology transfer also arises from an interest in promoting the economic growth that is vital to the nation's welfare and security. It is through further development, refinement, and marketing that the results of research become diffused throughout the economy and can generate growth. It is widely accepted that technological progress is responsible for up to one-half the growth of the U.S. economy and is the principal driving force in long-term economic growth and increases in our standard of living. Economic benefits of a technology or technique accrue when a product, process, or service is brought to the marketplace where it can be sold or used to increase productivity. When technology transfer is successful, new and different products or processes become available to meet or induce market demand. Transfer from the federal laboratories can result in substantial increases in employment and income generated at the firm level.

Cooperation with the private sector provides a means for federal scientists and engineers to obtain state-of-the-art technical information from the industrial community, which in various instances is more advanced than the government. Technology transfer is also a way to assist companies that have been dependent on defense contracts and procurement to convert to manufacturing for the civilian, commercial marketplace. Successful efforts range from advances in the commercial aviation industry, to the development of a new technology for use in advanced ceramics, to the development of the biotechnology sector.

Technology Transfer to State and Local Governments: Rationale for Federal Activity

The increasing demands on state and local governments to provide improved goods and services have been accompanied by a recognition that expanded technological expertise can help meet many of these needs. The transfer of technology and technical knowledge from government laboratories to state and local jurisdictions can allow for additional use of ideas and inventions that have been funded and created through federal R&D. Intergovernmental technology transfer can also help state and local officials meet responsibilities imposed by federal legislation.

As state and local governments increasingly look for technological solutions, the concept of "public technology" — the adaptation and utilization of new or existing technology to public sector needs — has emerged. The application of technology to State and local services is a complex and intricate procedure. In transferring technology from the federal laboratories, the application often can be direct. At other times, alterations in technical products and processes may be necessary for application in the state and local environment. However, this "adaptive engineering" generally is not extensive or expensive and can be accomplished by federal laboratory and state and local personnel working together.

State and local government concerns with regional economic growth also have focused attention on technology transfer as a mechanism to increase private sector innovation related activities within their jurisdiction. In order to develop and foster an entrepreneurial climate, many states and localities are undertaking the support of programs that assist high technology businesses, and that often use the federal laboratory system. State and local efforts to develop "incubator centers" for small companies may rely on cooperation with federal laboratories, which supply technical expertise to firms locating at the center. Other larger programs to promote innovation in the state, such as the Ben Franklin Partnership in Pennsylvania, use the science and technology resources of federal personnel. Additional programs have been created involving state universities, private companies, and the federal laboratories, with each program geared to the specific needs and desires of the participating parties. (For more discussion see CRS Report 98-859, *State Technology Development Strategies: The Role of High Tech Clusters*.)

Current Federal Efforts to Promote Technology Transfer

Over the years, several federal efforts have been undertaken to promote the transfer of technology from the federal government to state and local jurisdictions and to the private sector. The primary law affording access to the federal laboratory system is P.L. 96-480, the Stevenson-Wydler Technology Innovation Act of 1980, as amended by the Federal Technology Transfer Act of 1986 (P.L. 99-502), the Omnibus Trade and Competitiveness Act (P.L. 101-418), the 1990 Department of Defense (DOD) Authorization Act (P.L. 101-189), the National Defense Authorization Act for FY1991 (P.L. 101-510), the Technology Transfer Improvements and Advancement Act (P.L. 104-113), and the Technology Transfer

Commercialization Act (P.L. 106-404). Several practices have been established and laws enacted that are aimed at encouraging the private sector to utilize the knowledge and technologies generated by the federal R&D endeavor. These are discussed below.

Federal Laboratory Consortium for Technology Transfer

One of the primary federal efforts to facilitate and coordinate the transfer of technology among various levels of government and to the private sector is the Federal Laboratory Consortium for Technology Transfer (FLC). The Consortium was originally established under the auspices of the Department of Defense in the early 1970s to assist in transferring DOD technology to state and local governments. Several years later, it was expanded to include other federal departments in a voluntary organization of approximately 300 federal laboratories. The Federal Technology Transfer Act of 1986 (P.L. 99-502) provided the FLC with a legislative mandate to operate and required the membership of most federal laboratories. Today, over 600 laboratories are represented.

The basic mission of the Federal Laboratory Consortium is to promote the effective use of technical knowledge developed in federal departments and agencies by "networking" the various member laboratories with other federal entities, with state, local, and regional governments, and with private industry. To accomplish this, the Consortium establishes channels through which user needs can be identified and addressed. It also provides a means by which federal technology and expertise can be publicized and made available through individual laboratories to private industry for further development and commercialization. Access to the resources of the full federal laboratory system can be made through any laboratory representative, the FLC regional coordinators, the Washington area representative, or by contacting the Chairman or Executive Director.

The FLC itself does not transfer technology; it assists and improves the technology transfer efforts of the laboratories where the work is performed. In addition to developing methods to augment individual laboratory transfer efforts, the Consortium serves as a clearinghouse for requests for assistance and will refer to the appropriate laboratory or federal department. The work of the Consortium is funded by a set-aside of 0.008% of the portion of each agency's R&D budget used for the laboratories.

P.L. 96-480, P.L. 99-502, and Amendments

In 1980, the U.S. Congress enacted P.L. 96-480, the Stevenson-Wydler Technology Innovation Act. Recognizing the benefits to be derived from the transfer of technology, the law explicitly states that:

> It is the continuing responsibility of the federal government to ensure the full use of the results of the Nation's federal investment in research and development. To this end the federal government shall strive where appropriate to transfer federally owned or originated [non-classified] technology to state and local governments and to the private sector.

Prior to this law, technology transfer was not part of the mission requirements of the federal departments and agencies, with the exception of the National Aeronautics and Space Administration. This left laboratory personnel open to questions as to the suitability of their transfer activities. However, P.L. 96-480 "legitimized" the transfer effort and mandated that technology transfer be accomplished as an expressed part of each agency's mission.

Section 11 created the mechanisms by which federal agencies and their laboratories can transfer technology. Each department with at least one laboratory must make available not less than 0.5% of its R&D budget for transfer activities, although this requirement can and has been waived. To facilitate transfer from the laboratories, each one is required to establish an Office of Research and Technology Applications (ORTA); laboratories with annual budgets exceeding $20 million must have at least one full-time staff person for this office (although the latter provision can also be waived). The function of the ORTA is to identify technologies and ideas that have potential for application in other settings.

Additional incentives for the transfer and commercialization of technology are contained in various amendments to Stevenson-Wydler. P.L. 99-502, the Federal Technology Transfer Act, amends P.L. 96-480 to allow government-owned, government-operated laboratories (GOGOs) to enter into cooperative research and development agreements (CRADAs) with universities and the private sector. The authority to enter into these agreements was extended to government-owned, contractor-operated laboratories (generally the laboratories of the Department of Energy) in the FY1990 Defense Authorization Act (P.L. 101-189). A CRADA is a specific legal document (not a procurement contract) which defines the collaborative venture. It is intended to be developed at the laboratory level, with limited agency review. In agencies which operate their own laboratories, the laboratory director is permitted to make decisions to participate in CRADAs in an effort to decentralize and expedite the technology transfer process. Generally, at agencies which use contractors to run their laboratories, specifically DOE, the CRADA is to be approved by headquarters. P.L. 106-398, however, allows the agency to define certain conditions under which the CRADA may be approved by a laboratory itself rather than headquarters.

The work performed under a cooperative research and development agreement must be consistent with the laboratory's mission. In pursuing these joint efforts, the laboratory may accept funds, personnel, services, and property from the collaborating party and may provide personnel, services, and property to the participating organization. The government can cover overhead costs incurred in support of the CRADA, but is expressly prohibited from providing direct funding to the industrial partner. In GOGO laboratories, this support comes directly from budgeted R&D accounts. Prior to the elimination of a line item in the budget to support non-defense energy technology transfer, the Energy Department generally relied on a competitive selection process run by headquarters to allocate funding specifically designated to cover the federal portion of the CRADA. Now these efforts are to be supported through programmatic funds. A line item still exists for DOE defense program technology transfer, but at reduced levels from previous years.

Under a CRADA, title to, or licenses for, inventions made by a laboratory employee may be granted in advance to the participating company, university, or consortium by the director of the laboratory. In addition, the director can waive, in advance, any right of ownership the government might have on inventions resulting from the collaborative effort regardless of size of the company. This diverges from other patent law which only requires that title to inventions made under federal R&D funding be given to small businesses, not-for-profits, and universities. In all cases, the government retains a nonexclusive, nontransferable, irrevocable, paid-up license to practice, or have practiced, the invention for its own needs.

Laboratory personnel and former employees are permitted to participate in commercialization activities if these are consistent with the agencies' regulations and rules of conduct. Federal employees are subject to conflict of interest restraints. In the case of government-owned, contractor-operated laboratories, P.L. 101-189 required the establishment of conflict of interest provisions regarding CRADAs to be included in the laboratories' operating contracts within 150 days of enactment of the law. Preference for cooperative ventures is given to small businesses, companies which will manufacture in the United States, or foreign firms from countries that permit American companies to enter into similar arrangements. To date, over 5,000 CRADAs have been signed (including NASA Space Act Agreements).

P.L. 99-502 provides for cash awards to federal laboratory personnel for activities facilitating scientific or technological advancements which have either commercial value or contribute to the mission of the laboratory and for the transfer of technology leading to commercialization. As an additional incentive, federal employees responsible for an invention are to receive at least 15% of royalties generated by the licensing of the patent associated with their work. The agencies may establish their own royalty sharing programs within certain guidelines. If the government has the right to an invention but chooses not to patent, the inventor, either as a current or former federal employee, can obtain title subject to the above-mentioned licensing rights of the government.

To further facilitate the transfer process, a provision of the National Defense Authorization Act for FY1991 (P.L. 101-510) amends Stevenson-Wydler allowing government agencies and laboratories to develop partnership intermediary programs augmenting the transfer of laboratory technology to the small business sector.

P.L. 104-113, the Technology Transfer Improvements and Advancement Act, clarifies existing policy with respect to the dispensation of intellectual property under a CRADA by amending the Stevenson-Wydler Act. Responding to criticism that ownership of patents is an obstacle to the quick development of CRADAs, this bill guarantees an industrial partner the option to select, at the minimum, an exclusive license for a field of use to the resulting invention. If the invention is made solely by the private party, then they may receive the patent. However, the government maintains a right to have the invention utilized for compelling public health, safety, or regulatory reasons and the ability to license the patent should the industrial partner fail to commercialize the invention.

P.L. 100-418, Omnibus Trade and Competitiveness Act

In response to concerns over the development and application of new technology, the 1988 Omnibus Trade and Competitiveness Act contained several provisions designed to foster technology transfer. The law redesignated the National Bureau of Standards as the National Institute of Standards and Technology (NIST), and mandated the establishment of (among other things): (1) an Advanced Technology Program to encourage public-private cooperative efforts in the development of industrial technology and to promote the use of NIST technology and expertise; (2) Regional Manufacturing Technology Transfer Centers; and (3) a Clearinghouse on State and local innovation related activities. The set-aside for operation of the Federal Laboratory Consortium created in P.L. 99-502 was also increased from 0.005% of the laboratory R&D budget to 0.008%.

The Advanced Technology Program (ATP) provides seed funding, matched by private-sector investment, to companies or consortia of universities, industries, and government laboratories to accelerate the development of generic technologies that have broad application across industries. The first awards were made in 1991. As of May 2004, 736 projects have been funded representing approximately $2.2 billion in federal dollars matched by $2 billion in financing from the private sector.

The first four ATP competitions (through August 1994) were all general in nature. However, in response to large increases in federal funding, NIST, in conjunction with industry, identified various key areas for long-range support including information infrastructure for healthcare; tools for DNA diagnostics; component-based software; manufacturing composite structures; computer-integrated manufacturing for electronics; digital data storage; advanced vapor-compression refrigeration systems; motor vehicle manufacturing technology; materials processing for heavy manufacturing; catalysis and biocatalysis technologies; advanced manufacturing control systems; digital video in information networks; engineering; photonics manufacturing; premium power; microelectronics manufacturing infrastructure; selective-membrane platforms; and adaptive learning systems. The general competition continued. Since FY1999, NIST dropped the focused areas in favor of one competition open to all areas of technology. (For additional information see CRS Report 95-36, *The Advanced Technology Program*.)

Appropriations for the Advanced Technology Program included $36 million in FY1991, $48 million in FY1992, $67.9 million in FY1993, and $199.5 million in FY1994. Appropriations for FY1995 expanded significantly to $431 million, but $90 million of this amount was rescinded by P.L. 104-6 as funding for ATP met with opposition in the 104th Congress. The initial House Republican budget proposal associated with the House Republican "Contract with America" would have eliminated the Advanced Technology Program. In addition to rescinding $90 million from the FY1995 funding of ATP, both the House and Senate failed to authorize spending for this activity. The appropriations bill that originally passed Congress, H.R. 2076, was vetoed by the President, in part, because it offered no support for ATP. The legislation that was finally signed into law, P.L. 104-134, funded the program at $221 million. In the last session of the 104th Congress, again there were no FY1997 authorizations for the Advanced Technology Program. However, the Omnibus Consolidated Appropriations Act, P.L. 104-208, provided $225 million in FY1997 financing for ATP. P.L. 105-18 rescinded $7 million from this amount. President Clinton requested $276 million for ATP in his FY1998 budget. While no authorization legislation was enacted, P.L. 105-119 appropriated $192.5 million for ATP in FY1998.

The Administration's FY1999 budget proposed $259.9 million for ATP, an increase of 35%. While not providing such a large increase, P.L. 105-277, the Omnibus Consolidated Appropriations Act, does fund ATP for FY1999 at $197.5 million, 3% above the previous year. This reflects a $6 million rescission of "deobligated" funds resulting from early termination of certain projects. For FY2000, the Administration proposed support for ATP at $238.7 million, an increase of 21% above last year's funding. No authorization legislation was enacted. S. 1217, passed by the Senate on July 22, 1999, would have appropriated $226.5 million for ATP, 15% more than the current year. In contrast, the appropriations bill passed by the House on August 5, 1999, H.R. 2670, contained no funding for the Advanced

Technology Program. The report to accompany the bill stated that "...the program has not produced a body of evidence to overcome those fundamental questions about whether the program should exist in the first place." Yet P.L. 106-113, the Consolidated Appropriations Act, provided $142.6 million for ATP, although this represented a 28% decrease over FY1999.

President Clinton's FY2001 budget called for financing ATP at $175.5 million, 23% above the prior fiscal year. The original version of the appropriations bill passed by the House did not fund the program. However, P.L. 106-553 appropriated $145.7 million for ATP, an increase of 2% from the previous funding level.

For FY2002, the Bush Administration's budget proposal would have suspended support for all new awards pending an evaluation of the program; $13 million would be made available to meet financial commitments for on-going projects. H.R. 2500, as initially passed by the House, contained no funding for new ATP grants but also provided $13 million to support prior project commitments. The original version of H.R. 2500 passed by the Senate funded ATP at $204.3 million. The final legislation, P.L. 107-77, financed the program at $184.5 million, an increase of almost 27% over the previous fiscal year.

In the FY2003 budget request, the Advanced Technology Program would have received $108 million, 35% below the FY2002 figure. No relevant appropriations legislation was enacted in the 107th Congress. A series of Continuing Resolutions provided support for ATP at the FY2002 level until the 108th Congress passed P.L. 108-7 which appropriated $178.8 million for the program in FY2003 (after the 0.65% across the board recision mandated by the legislation).

The President's FY2004 budget would have provided $27 million for ATP to cover ongoing commitments; no new projects would be funded. H.R. 2799, as initially passed by the House, appropriated no money for ATP. However, H.R. 2799 was incorporated into H.R. 2673, which became P.L. 108-199, the FY2004 Consolidated Appropriations Act. Signed into law on January 23, 2004, this legislation provides $170.5 million for ATP after a rescission mandated in the bill.

For FY2005, the Administration budget request and H.R. 4754, the appropriations bill originally passed by the House on July 8, 2004, did not include funding for ATP. S. 2809, as reported to the Senate by the Committee on Appropriations, would have provided $203 million for the program. However, P.L. 108-447, the FY2005 Omnibus Appropriations Act, funds ATP at $136.5 million (after several rescissions mandated in the legislation), a 20% decrease from the previous fiscal year.

The Administration's FY2006 budget, and H.R. 2862, as passed by the House do not include support for the Advanced Technology Program. However, the version of H.R. 2862 reported to the Senate from the Committee on Appropriations would provide $140 million for the program.

The Omnibus Trade and Competitiveness Act (P.L. 100-418) also created a program of regional centers to assist small manufacturing companies' use of knowledge and technology developed under the auspices of the National Institute of Standards and Technology and other federal agencies. Federal funding for the centers is matched by non-federal sources including state and local governments and industry. Originally, seven Regional Centers for the Transfer of Manufacturing Technology were selected and are operational: the Great Lakes Manufacturing Technology Center at the Cleveland Advanced Manufacturing Program in Ohio; the Northeast Manufacturing Technology Center at Rensselaer Polytechnic Institute in

Troy, New York (now called the New York Manufacturing Extension Partnership); the South Carolina Technology Transfer Cooperative based at the University of South Carolina in Columbia; the Midwest Manufacturing Technology Center at the Industrial Technology Institute in Ann Arbor, Michigan; the Mid-American Manufacturing Technology Center at the Kansas Technology Enterprise Corporation of Topeka; the California Manufacturing Technology Center at El Camino College in Torrance; and the Upper Midwest Manufacturing Technology Center in Minneapolis.

The original program expanded in 1994 creating the Manufacturing Extension Partnership (MEP) to meet new and growing needs of the community. In a more varied approach, the Partnership involves both large centers and smaller, more dispersed organizations sometimes affiliated with larger centers. Also included is the NIST State Technology Extension Program which provides states with grants to develop the infrastructure necessary to transfer technology from the federal government to the private sector (an effort which was also mandated by P.L. 100-418) and a program that electronically ties the disparate parties together along with other federal, state, local, and academic technology transfer organizations. There are now centers in all 50 states and Puerto Rico. Since the program was created in 1989, awards made by NIST for extension activities resulting in the creation of approximately 400 regional offices. [It should be noted that the Department of Defense also funded 36 centers through its Technology Reinvestment Project (TRP) in FY1994 and FY1995. When the TRP was terminated, NIST took over support for 20 of these programs in FY1996 and financed the remaining ones during FY1997.]

Funding for this program was $11.9 million in FY1991, $15.1 million in FY1992, and $16.9 million in FY1993. The FY1994 appropriation for the expanded Manufacturing Technology Partnerships was $30.3 million. P.L. 103-317 appropriated $90.6 million for this effort in FY1995, although P.L. 104-19 rescinded $16.3 million from this appropriation. For FY1996, H.R. 2076, which passed the Congress but was vetoed by the President, included appropriations of $80 million for MEP. This amount was retained in the final legislation, P.L. 104-134. The President's FY1997 budget request for this program was $105 million. No FY1997 authorization legislation was enacted, but P.L. 104-208 appropriated $95 million for Manufacturing Extension while temporarily lifting the six-year limit on federal support for individual centers. The Administration's MEP budget request for FY1998 was $123 million. Again, no authorizing legislation was enacted but P.L. 105-119 provided $113.5 million in appropriations for FY1998. The law also permitted government funding for the centers to be extended, for periods of one year and at a rate of one-third the centers annual costs if a positive evaluation was received. The President requested MEP funding of $106.8 million for FY1999, a decrease of 6%. P.L. 105-277 appropriated the $106.8 million. The decrease in funding reflects a reduced federal financial commitment as the centers mature, not a decrease in program support. In addition, the Technology Administration Act of 1998, P.L. 105-309, permits the federal government to fund centers at one-third the cost after the six years if a positive independent evaluation is made every two years. For FY2000, the Administration proposed support for MEP at $99.8 million. There were no authorizations enacted. S. 1217, as passed by the Senate, would have appropriated $109.8 million for the program, an increase of 3% over FY1999. H.R. 2670, as passed by the House, would have provided an appropriation of $99.8 million, as per the President's request. The version of H.R. 2670, as passed by both the House and Senate, appropriated $104.8 million; however, this legislation was vetoed by

President Clinton. P.L. 106-113, the Consolidated Appropriations Act, provided $104.2 million (after the mandated rescission).

The FY2001 Clinton budget requested $114.1 million in MEP funding, almost 9% above the previous fiscal year. The increase was designated for a new e-commerce outreach effort with the Department of Agriculture and the Small Business Administration. P.L. 106-553 appropriates $105.1 million for the Manufacturing Extension Partnership but does not permit the creation of any new programs.

The Bush Administration's FY2002 budget proposal included funding of $106.3 million for MEP. The version of H.R. 2500 first passed by the House would provide $106.5 million for the program; the original Senate-passed version would fund the activity at $105.1 million. P.L. 107-77 provides $106.5 million.

The FY2003 budget request included an 89% decrease in funding for MEP to $13 million. According to the budget document, "...consistent with the program's original design, the President's budget recommends that all centers with more than six years experience operate without federal contribution." A series of Continuing Resolutions financed the activity at the FY2002 level until the 108th Congress enacted P.L. 108-7 which appropriates $105.9 million for FY2003 (after the mandated recision).

The President's FY2004 budget proposed funding of $12.6 million for manufacturing extension to support those centers that have not reached six years of federal financing. H.R. 2799, the FY2004 appropriations bill, as initially passed by the House, financed the Partnership at $39.6 million. This bill was incorporated into H.R. 2673, which was signed into law as P.L. 108-199, the FY2004 Consolidated Appropriations Act. The legislation funds MEP at $38.7 million after a mandated rescission contained in the bill.

For FY2005, the Bush Administration's budget request included $39.2 million for the MEP. H.R. 4754, the Commerce, Justice, State appropriations bill originally passed by the House would finance manufacturing extension at $106 million. S. 2809, reported to the Senate by the Committee on Appropriations, would have provided $112 million to "fully fund" existing centers and to provide additional assistance to small and rural States. The FY2005 Omnibus Appropriations Act, P.L. 108-447, restores earlier cuts to the program and finances MEP at $107.5 million after several rescissions mandated by the legislation.

In the President's FY2006 budget proposal, support for manufacturing extension would be reduced to $46.8 million, 63% below the current fiscal year. H.R. 2862, as passed by the House and as reported to the Senate from the Committee on Appropriations, would fund the program at $106 million. (For additional information see CRS Report 97-104, *Manufacturing Extension Partnership Program: An Overview.*)

Patents

The patent system was created to promote innovation. Based on Article I, Section 8 of the U.S. Constitution which states: "The Congress Shall Have Power... To promote the Progress of Science and useful Arts, by securing for limited Times to Authors and Inventors the exclusive Right to their respective Writings and Discoveries...", the patent system encourages innovation by simultaneously protecting the inventor and fostering competition. Originally, it provided the inventor with a lead time of 17 years (from the date of issuance) to develop his idea, commercialize, and thereby realize a return on his initial investment. Today, in response to the Uruguay Round Agreements Act, the term of the patent has been changed to 20 years from date of filing. The process of obtaining a patent places the idea in the public domain. As a disclosure system, the patent can, and generally does, stimulate other firms or inventors to invent "around" existing patents to provide parallel technical developments or meet similar market needs.

Ownership of patents derived from research and development performed under federal funding affects the transfer of technology from federal laboratories to the private sector. Generally, the government retains title to these inventions and can issue to companies either an exclusive license or, more commonly, a nonexclusive license. However, it is argued that without title (or at least an exclusive license) to an invention and the protection it conveys, a company will not invest the additional time and money necessary for commercialization. This contention is supported by the fact that, although a portion of ideas patented by the federal government have potential for further development, application, and marketing, only about 10% of these are ever used in the private sector. However, there is no universal agreement on this issue. It also is asserted that title should remain in the public sector where it is accessible to all interested parties since federal funds were used to finance the work.

Despite the disagreements, the Congress has accepted to some extent the proposition that vesting title to the contractor will encourage commercialization. P.L. 96-517, Amendments to the Patent and Trademark Laws (commonly known as the Bayh-Doyle Act), provides, in part, for contractors to obtain title if they are small businesses, universities, or not-for-profit institutions. Certain rights are reserved for the government and these organizations are required to commercialize within a predetermined and agreed-on time. (For more information see CRS Report RL32076, *The Bayh-Dole Act: Selected Issues in Patent Policy and the Commercialization of Technology*, CRS Report 98-862, *R&D Partnerships and Intellectual Property: Implications for U.S. Policy* and CRS Report RL30320, *Patent Ownership and Federal Research and Development: A Discussion on the Bayh-Dole Act and the Stevenson-Wydler Act.*) Yet it continues to be argued that patent exclusivity is important for both large and small firms. In a February 1983 memorandum concerning the vesting of title to inventions made under federal funding, President Reagan ordered all agencies to treat, as allowable by law, all contractors regardless of size as prescribed in P.L. 96-517. This, however, does not have a legislative basis.

Further changes in the patent laws made by the enactment of P.L. 98-620 also affect the transfer of technology from federal laboratories to the private sector. In a provision that was designed to increase interaction and cooperation between government-owned, contractor-operated (GOCO) laboratories and private industry in the transfer of technology, Title V permits decisions to be made at the laboratory level as to the award of licenses for laboratory generated patents. The contractor is also permitted by this legislation to receive patent royalties for use in additional research and development, for awards to individual inventors on staff, or for education. A cap exists on the amount of the royalty returning to the laboratory so as not to distort the agency's mission and congressionally mandated R&D agenda. However, the creation of discretionary funds gives laboratory personnel added incentive to encourage and complete technology transfers.

P.L. 98-620 also permits private companies, regardless of size, to obtain exclusive license for the full life of the government patent. Prior restrictions on large firms allowed exclusive license for only 5 of the (then) 17 years of the patent. The law permits those government laboratories that are run by universities or nonprofit institutions to retain title to inventions made in their institution (within certain defined limitations). Federal laboratories operated by large companies are not included in this provision.

The Federal Technology Transfer Act and the FY1990 DOD authorization gives all companies (not just small businesses, universities, and nonprofits) the right to retain title to inventions resulting from research performed under cooperative R&D agreements with government laboratories. If this occurs, the federal government retains a royalty-free license to use these patents. In addition, the Federal Technology Act states that the government agencies may retain a portion of royalty income rather than returning it to the Treasury. After payment of the prescribed amount to the inventor, the agencies must transfer the balance of the total to their government-operated laboratories, with the major portion distributed to the laboratory where the invention was made. The laboratory may keep all royalties up to 5% of their annual budget plus 25% of funds in excess of the 5% limit. The remaining 75% of the excess returns to the Treasury. Funds retained by the laboratory are to be used for expenses incurred in the administration and licensing of inventions; to reward laboratory personnel; to provide for personnel exchanges between laboratories; for education and training consistent with the laboratories' and agencies' missions; or for additional transfer.

P.L. 106-404, the Technology Transfer Commercialization Act, signed into law on November 1, 2000, makes alterations in current practices concerning patents held by the government to make it easier for federal agencies to license such inventions. The law amends P.L. 98-480 and P.L. 96-517 to decrease the time necessary to obtain an exclusive or partially exclusive license on federally owned patents. Previously, agencies were required to publicize the availability of technologies for three months using the *Federal Register* and then provide an additional 60 day notice of intent to license by an interested company. The new law shortens the period to 15 days in recognition of the ability of the Internet to offer widespread notification and the time constraints faced by industry in commercialization activities. Certain rights would be retained by the government. The legislation also allows licenses for existing government-owned inventions to be included in CRADAs.

The CREATE Act, P.L. 108-453, makes changes in the patent laws to promote cooperative research and development among universities, government, and the private sector. The legislation amends section 103(c) of title 25, United States Code, such that certain actions between researchers under a joint research agreement will not preclude patentability.

Small Business Technology Transfer Program

P.L. 102-564 created a three-year pilot program designed to facilitate the commercialization of university, nonprofit, and federal laboratory R&D by small companies. The Small Business Technology Transfer program (STTR) provides funding for research proposals which are developed and executed cooperatively between a small firm and a scientist in a research organization and fall under the mission requirements of the federal funding agency. Up to $100,000 in Phase I financing is available for one year to test the viability of the concept. Phase II awards of $500,000 may be made for two years to perform the research. Funding for commercialization of the results is expected from the private sector. Financial support for this effort comes from a phased-in set-aside of the R&D budgets of departments which spend over $1 billion per year on research and development. Originally set to expire at the end of FY1996, the program was extended one year. P.L. 105-135 reauthorizes funding through FY2001. In the current Congress, P.L. 107-50 extends the STTR program through FY2009. The activity will be funded by an increase in the set-aside to 0.3% beginning in FY2004. Also in FY2004, the amount of money available for individual Phase II grants will increase from $500,000 to $750,000. (For additional information see CRS Report 96-402, *Small Business Innovation Research Program*.)

FURTHER CONSIDERATIONS

The federal laboratories have received a mandate to transfer technology. This, however, is not the same as a mandate to help the private sector in the development and commercialization of technology for the marketplace. While the missions of the government laboratories are often broad, direct assistance to industry is not included, with the exception of the National Institute of Standards and Technology. The laboratories were created to perform the R&D necessary to meet government needs, which typically are not consistent with the demands of the marketplace.

The missions of the federal laboratories are under review, due, in part, to budget constraints and the changing world order. National security is now being redefined to include economic well-being in addition to weapons superiority. The laboratories which have contributed so much to the defense enterprise are being re-evaluated. These discussions provide an opportunity to debate whether the mandate of the federal R&D establishment should include expanded responsibilities for assistance to the private sector. Whether or not the missions of the U.S. government laboratories are changed to include expanded assistance to industry, there are various initiatives which may facilitate the technology transfer process under the laboratories' current responsibilities. These include making the work performed in government institutions more relevant to industry through augmented cooperative R&D, increased private sector involvement early in the R&D efforts of the laboratories, and expanded commercialization activities.

Because a significant portion of the laboratories are involved in defense research, questions arise as to whether or not the technologies in these institutions can be transferred in such a way as to be useful to commercial companies. In addition, the selection of one company over another to be involved in a transfer or in a cooperative R&D agreement raises issues of fairness and equity of access, as well as conflict of interest. And, while it is virtually impossible to prevent the flow of scientific and technical information abroad, there is ongoing interest in the extent of foreign access to the federal laboratory establishment. How these concerns are addressed may be fundamental to the success of U.S. technology transfer.

Over the past 25 years, the Congress has enacted various laws designed to facilitate cooperative R&D between and among government, industry, and academia. These laws include (but are not limited to) tax credits for industrial payments to universities for the performance of R&D, changes in the antitrust laws as they pertain to cooperative research and joint manufacturing, and improved technology transfer from federal laboratories to the private sector. The intent behind these legislative initiatives is to encourage collaborative ventures and thereby reduce the risks and costs associated with R&D as well as permit work to be undertaken that crosses traditional boundaries of expertise and experience leading to the development of new technologies and manufacturing processes for the marketplace.

Since the 104th Congress, the perspectives on joint R&D, technology transfer, and cooperative research and development agreements appear mixed. The results of legislative activity are open to discussion. In the recent past, both national political parties have supported measures to facilitate technological advancement. There are indications that the congressional majority favors refocusing federal support for basic research as well as indirect measures to encourage technology development in the private sector. CRADAs, in particular, are a means to take this government-funded basic research from the federal laboratory system and move it to the industrial community for commercialization to meet both agency mission requirements and other national needs associated with the economic growth which comes from new products and processes. It should also be recognized that the government is expressly prohibited from providing direct financial support to partners in the cooperative venture under a CRADA. Thus, it appears that this approach may meet the criteria expressed as acceptable to the Congress. While the Advanced Technology Program has faced much opposition in the House, the program continues to be funded, although at decreased levels. Recently, the Manufacturing Extension Partnership had its budget cut, but these cuts were restored the following fiscal year with the support of the Congress. As the 109th Congress begins to make decisions concerning funding for R&D, the role of the federal government in technology transfer, technology development, and commercialization might be expected to be explored further.

LEGISLATION

H.R. 250 (Ehlers)

Manufacturing Technology Competitiveness Act. Creates an interagency committee to coordinate federal manufacturing R&D. Establishes and authorizes funding for a pilot collaborative manufacturing research grants program to promote the development of new manufacturing technologies through cooperative applied research among the private sector, academia, states, and other non-profit institutions. Mandates and authorizes financing for a manufacturing fellowship program. Creates and authorizes support for a manufacturing extension center competitive grants program to focus on new or emerging manufacturing technologies. Authorizes funding for the Manufacturing Extension Partnership, among other things. Introduced January 6, 2005; referred to the Committee on Science. Reported to the House, amended, May 23, 2005.

H.R. 1454 (Sensenbrenner)

Amends the Internal Revenue Code of 1986 to make the research and experimentation tax credit permanent. Introduced March 20, 2005; referred to the House Committee on Ways and Means.

H.R. 1736 (Johnson, N.)

Amends the Internal Revenue Code of 1986 to make the research and experimentation tax credit permanent, among other things. Introduced April 20, 2005; referred to the House Committee on Ways and Means.

H.R. 2862 (Wolf)

Makes appropriations for science and the Departments of State, Justice, and Commerce. As passed by the House, the bill would provide $106 million for the Manufacturing Extension Partnership and no financing for the Advanced Technology Program. The version of the legislation reported to the Senate from the Committee on Appropriations would fund MEP at $106 million and provide $140 for ATP. Introduced June 10, 2005; referred to the House Committee on Appropriations. Passed the House, amended, on June 16, 2005. Received in the Senate on June 16, 2005; referred to the Senate Committee on Appropriations. Reported to the Senate, with an amendment in the nature of a substitute, on June 23, 2005.

S. 296 (Kohl)

Authorizes appropriations for the Manufacturing Extension Partnership through FY2008, including $115 million for FY2006. Introduced February 3, 2005; referred to the Senate Committee on Commerce, Science, and Transportation.

S. 1020 (Coleman)

"COMPETE" Act of 2005. Extends the research and experimentation tax credit through the end of 2007, among other things. Introduced May 12, 2005; referred to the Senate Committee on Finance.

INDEX

A

academics, 29
access, xi, 8, 9, 20, 28, 40, 65, 68, 108, 110, 111, 120, 121, 123, 128, 132, 141, 151
accountability, 40
accounting, 61
achievement, 99
acid, 98
adaptation, 55, 141
additives, 98
adhesion, 98
advertising, 49, 61, 110, 111, 116
aerospace, 132
affect, 6, 40, 66, 68, 82, 104, 121, 149
agent, 46, 56
alloys, 43
alternative(s), 4, 7, 8, 20, 25, 28, 64, 83, 127
alters, 89
amendments, x, 29, 88, 119, 143
America Online, 116
amino acids, 98
angiogenesis, 99
animals, 45, 97, 99, 101
antibiotic, 98
antitrust, 67, 78, 87, 91, 120, 121, 124, 133, 138, 151
antitumor, 45
applied research, 79, 122, 134, 152
arginine, 98
argument, 55, 64, 68, 101, 109, 111, 114
assessment, 26, 32, 68
assets, 4
assignment, 52, 139
assumptions, 81
asymmetry, 8
ATP, 78, 84, 85, 93, 120, 124, 125, 126, 134, 138, 145, 146, 153
attacks, 115
attention, viii, 2, 39, 60, 66, 141
authority, 3, 11, 16, 18, 23, 34, 49, 88, 143
availability, viii, xi, 4, 11, 20, 24, 39, 40, 41, 56, 60, 61, 62, 89, 97, 120, 123, 130, 131, 150
awareness, 87

B

bargaining, 8
barriers, 82, 124, 139
basic research, 78, 79, 82, 83, 85, 86, 98, 101, 104, 120, 122, 123, 126, 133, 139, 152
behavior, 4, 6, 8, 12, 25
beneficial effect, 44
biocompatibility, 98
biological activity, 98
biotechnology, 8, 40, 66, 79, 100, 102, 122, 140
blindness, 114
blood, 54, 99
blood vessels, 99
bloodstream, 105
body, 82, 126, 146
brand loyalty, 62
Brazil, 57
business model, 9

C

candidates, 99
cast, 114
catheter, 24
cell, 45, 98, 102
cell line, 45, 102
cell surface, 98
certificate, 28, 47
channels, 142
chicken, 99

children, 60, 66
clients, 112
clinical trials, 44, 53, 54, 60, 67, 99
cloning, 102
collaboration, 81, 99, 123, 128
commitment, 81, 90, 129, 132, 147
communication, 111
community, vii, xi, 1, 2, 5, 9, 11, 19, 44, 45, 81, 83, 85, 86, 87, 89, 121, 122, 124, 127, 128, 131, 137, 139, 140, 147, 152
comparative advantage, 79, 121
compensation, 4, 6, 14, 40, 41, 45, 52, 63
competition, x, 8, 41, 58, 62, 65, 78, 81, 91, 119, 121, 124, 125, 131, 145, 148
competitiveness, vii, viii, x, 42, 77, 80, 81, 86, 119, 121, 122, 128, 129, 130, 131, 132
competitor, 21, 28, 54
complexity, vii, 1, 5, 11, 16
compliance, 54, 57
components, 2, 46, 80, 81
composition, 43
compounds, ix, 43, 44, 45, 52, 95, 96, 97, 98, 100, 101, 102, 103
computer software, 117
computer use, 115
concentration, 5, 122
conception, 122
conduct, vii, 1, 2, 4, 7, 14, 17, 18, 27, 46, 60, 96, 97, 99, 103, 104, 108, 109, 110, 144
confidence, 34
conflict, 12, 32, 55, 56, 67, 144, 151
conflict of interest, 144, 151
conformity, 101
confusion, 98, 103, 116
Congressional Budget Office, 61, 62, 64, 65, 74
consensus, viii, 77, 79, 80
consent, 20
construction, 96, 101, 102, 103
consumers, ix, 53, 54, 62, 64, 95, 97, 113, 115, 130
control, x, 41, 99, 108, 110, 113, 145
corporations, 23, 32, 89
costs, x, 2, 5, 6, 7, 16, 19, 33, 43, 54, 56, 60, 64, 66, 67, 83, 86, 119, 121, 130, 131, 139, 143, 147, 151
costs of production, 131
counsel, 27
Court of Appeals, ix, 3, 24, 27, 45, 48, 51, 54, 55, 57, 58, 95, 96, 100, 104, 107, 110, 111, 112, 114
coverage, 103
credit, 83, 86, 120, 126, 127, 138
criticism, 5, 16, 144
curing, 105
curiosity, 49
customers, 64, 109, 114, 116

D

damage, 87, 124
database, 112
decision making, 20
decisions, ix, xi, 6, 27, 48, 51, 80, 81, 88, 107, 114, 115, 120, 129, 130, 139, 143, 149, 152
deduction, 86, 127
defendants, x, 108, 109, 110, 111, 113, 115
defense, x, xi, 16, 17, 18, 21, 22, 23, 35, 36, 49, 54, 55, 58, 61, 67, 97, 99, 107, 113, 114, 115, 120, 123, 127, 131, 132, 139, 140, 143, 151
deficiency, 24
definition, 116, 132
delivery, 130
demand, xi, 45, 62, 132, 137, 140
denial, 97
Department of Agriculture, 90, 148
Department of Defense, 87, 88, 89, 127, 128, 131, 141, 142, 147
Department of Energy, 128, 131, 132, 143
Department of Justice, 87
derivatives, 99
desire, 5, 7, 22, 27, 44, 62, 104
development policy, 82
differentiation, 79, 122
diffusion, 79, 130
digital technologies, 115
direct measure, 62, 78, 91
disclosure, 4, 8, 16, 18, 25, 42, 87, 121, 124, 149
disseminate, 88
distribution, 79, 109, 112, 117, 121
divergence, 81
DNA, 102, 125, 145
doctors, 62
domain, 3, 4, 13, 41, 42, 44, 45, 53, 109, 116, 117, 149
donations, 86, 127
doors, 113
dosage, 54, 63
drugs, viii, ix, 39, 40, 52, 53, 54, 56, 57, 59, 60, 61, 62, 63, 64, 65, 66, 67, 71, 95, 96, 97, 98, 100, 101, 102, 103, 104, 139
duplication, 33, 43
duration, 21

E

East Asia, 130
e-commerce, 90, 148
economic development, 129

economic growth, vii, viii, xi, 1, 4, 8, 41, 77, 78, 79, 80, 81, 82, 121, 123, 124, 137, 139, 140, 141, 152
economic policy, 81
educational process, 85, 123
elderly, 40
employees, 15, 88, 144
employment, 140
encouragement, 7
energy, 82, 140, 143
England, 37
entrepreneurs, 6, 8, 32
environment, viii, 4, 10, 40, 42, 65, 66, 68, 81, 91, 122, 123, 140, 141
equipment, x, 83, 86, 102, 113, 119, 123, 126, 127, 131
equity, 23, 24, 25, 51, 67, 151
ester, 56
Europe, vii, 2, 12, 13, 24, 29
evidence, 24, 45, 46, 64, 84, 99, 100, 109, 113, 117, 126, 146
exercise, 27, 56, 111
expectation, 6, 10
expenditures, 9, 32, 60, 66, 83, 86, 126, 139
expertise, x, xi, 8, 25, 28, 81, 87, 88, 119, 121, 122, 123, 131, 132, 137, 139, 140, 141, 142, 144, 151
experts, vii, viii, 1, 2, 40, 42, 61, 62, 63, 64, 65, 66, 67
exports, 51

F

failure, 16, 18, 24, 27, 45, 96, 102
fair use, 97, 113
fairness, vii, 1, 2, 23, 151
faith, 17, 18, 27, 36, 87
FDA, viii, ix, 39, 52, 53, 54, 55, 56, 57, 58, 59, 60, 61, 63, 64, 65, 66, 67, 68, 72, 75, 76, 95, 96, 97, 98, 99, 100, 101, 102, 103, 104, 106
FDA approval, ix, 53, 54, 55, 56, 57, 63, 66, 68, 95, 97, 98, 100, 101, 102, 103
fear, 27, 53, 80
federal courts, 3, 98
federal funds, 79, 129, 130, 149
federal law, 40, 53
finance, 67, 84, 90, 148, 149
financial resources, 9
financial support, 84, 121, 123, 131, 152
financing, 41, 84, 120, 124, 125, 126, 132, 133, 134, 138, 145, 146, 148, 150, 152, 153
firms, 5, 6, 7, 8, 9, 10, 11, 19, 23, 25, 27, 29, 32, 42, 61, 62, 64, 67, 78, 79, 83, 84, 89, 102, 120, 121, 122, 124, 125, 128, 129, 131, 132, 138, 139, 141, 149

flexibility, 121
flood, 15
food, 44, 50, 53, 55, 98
food additives, 98
foreign firms, 60, 80, 121, 132, 144
foreign person, 87, 124
free riders, 4
funding, ix, xi, 9, 60, 62, 77, 78, 82, 83, 84, 85, 89, 90, 91, 99, 119, 120, 121, 122, 123, 124, 126, 130, 131, 133, 134, 138, 143, 145, 146, 147, 148, 149, 150, 152

G

gene, 9
General Motors, 71
generation, x, 79, 100, 115, 119, 128
generic drugs, 55, 57, 62, 65, 68, 103
Germany, 23
glycine, 98
goals, viii, xi, 4, 16, 27, 40, 42, 66, 68, 82, 91, 112, 123, 137
goods and services, xi, 4, 6, 32, 41, 80, 123, 130, 131, 137, 139, 140, 141
government, ix, x, xi, 9, 40, 61, 77, 78, 79, 80, 81, 82, 84, 85, 86, 88, 89, 90, 91, 117, 119, 120, 121, 122, 123, 124, 127, 128, 129, 130, 131, 133, 137, 138, 139, 140, 141, 142, 143, 144, 145, 147, 149, 150, 151, 152
government intervention, 80
government policy, 81, 130
grants, 2, 15, 18, 22, 47, 54, 60, 73, 83, 85, 89, 125, 129, 134, 146, 147, 151, 152
groups, 81, 112
growth, vii, viii, 4, 9, 32, 41, 60, 66, 67, 77, 79, 81, 82, 102, 140
growth factor, 102
growth rate, 66
guidelines, 67, 121, 130, 144
guilty, 49

H

hands, 110
harassment, 8
harm, 24, 51
harmonization, vii, 1, 2, 5, 7
health, 24, 61, 82, 123
higher education, 86, 127
hiring, 121
host, 113

House, 78, 83, 84, 85, 90, 91, 92, 93, 120, 126, 133, 134, 135, 138, 145, 146, 147, 148, 152, 153
human subjects, 98
hunting, 45
hypertension, 50

I

ideas, 10, 42, 81, 87, 88, 128, 141, 143, 149
imitation, 43
immunity, 114
implementation, 16, 40, 58, 60, 63, 64, 65, 66, 68, 88, 99
imports, 49
in vitro, 45, 99
incentives, viii, xi, 2, 5, 6, 25, 29, 40, 41, 42, 44, 50, 52, 60, 65, 66, 68, 78, 81, 82, 88, 91, 92, 124, 128, 131, 133, 137, 139, 143
income, 80, 122, 140, 150
increased competition, 65
independence, 53
indication, 63, 129
indicators, 62
indirect measure, 78, 91, 120, 138, 152
industrial policy, 80, 81
industrial sectors, 43, 61, 79, 121, 123, 124, 131
industrialized countries, 81
industry, viii, ix, x, 2, 5, 6, 8, 9, 10, 21, 23, 29, 39, 40, 41, 43, 48, 49, 50, 52, 53, 60, 61, 62, 65, 66, 77, 78, 79, 80, 81, 82, 85, 86, 87, 88, 89, 91, 92, 114, 119, 120, 121, 122, 123, 124, 126, 127, 128, 129, 130, 131, 132, 138, 139, 140, 142, 145, 146, 149, 150, 151
inelastic, 62
infancy, 66
influence, viii, 62, 77
infrastructure, 82, 89, 125, 145, 147
innovation, vii, viii, x, 1, 2, 4, 5, 6, 8, 11, 23, 26, 29, 39, 40, 41, 42, 60, 62, 63, 65, 66, 67, 68, 78, 79, 81, 82, 85, 91, 119, 121, 122, 123, 124, 129, 130, 139, 141, 144, 148
institutions, 9, 29, 83, 86, 89, 123, 125, 126, 128, 134, 149, 151, 152
instruments, 22
integration, 80
integrin, 99
integrity, 22
intellectual property, vii, 1, 7, 8, 15, 25, 40, 41, 66, 78, 88, 120, 128, 133, 138, 144
intellectual property rights, 40, 78, 88, 120, 133, 138
intensity, viii, 40, 62, 65, 66
intent, ix, xi, 14, 50, 58, 63, 64, 67, 68, 89, 96, 98, 101, 103, 128, 137, 150, 151

intentions, 17
interaction, 86, 88, 127, 132, 139, 140, 149
interest, vii, viii, xi, 1, 2, 4, 8, 9, 15, 23, 24, 39, 40, 43, 44, 45, 46, 48, 52, 54, 60, 65, 77, 80, 83, 100, 106, 110, 123, 124, 129, 131, 137, 139, 140, 144, 151
interface, 53
interference, 11, 47, 49, 81
international standards, 7
international trade, 79, 121
internationalization, 81
internet, 89
interpretation, 57, 96, 100, 101, 103
interrelationships, 80
inventions, vii, ix, xi, 1, 4, 5, 6, 9, 10, 11, 12, 14, 16, 21, 23, 25, 26, 40, 43, 44, 48, 50, 78, 86, 88, 89, 95, 96, 97, 98, 100, 101, 102, 103, 119, 122, 127, 128, 141, 143, 149, 150
investment, viii, 4, 5, 6, 23, 40, 41, 42, 60, 61, 65, 82, 88, 91, 142, 145, 148
investors, 6
isolation, 85
Israel, 74

J

Japan, vii, 2, 12, 13, 14, 23, 29, 130
job creation, 82
jobs, 8, 79, 121
judges, 29, 46, 48
judgment, x, 18, 23, 25, 59, 99, 100, 107, 108, 109, 113, 115
jurisdiction, 3, 7, 80, 141

K

knowledge, x, 2, 3, 4, 9, 17, 27, 41, 42, 44, 53, 81, 87, 89, 109, 110, 111, 114, 119, 122, 123, 128, 132, 139, 140, 141, 142, 146
knowledge-based economy, 2

L

labor, 79
language, 11, 13, 23, 24, 26, 49, 55, 98, 102, 116
laws, vii, viii, x, 1, 5, 7, 10, 13, 14, 17, 26, 39, 52, 53, 55, 66, 67, 78, 86, 87, 91, 92, 119, 120, 124, 126, 127, 128, 133, 138, 142, 149, 150, 151
lead, ix, 4, 5, 6, 10, 11, 32, 42, 43, 44, 95, 98, 101, 122, 131, 148
leadership, 91, 130
learning, 32, 42, 43, 145

legislation, vii, viii, 1, 2, 5, 6, 7, 8, 9, 10, 12, 13, 14, 15, 18, 19, 20, 21, 23, 24, 26, 27, 28, 29, 40, 59, 60, 61, 62, 63, 65, 66, 75, 78, 82, 84, 85, 86, 87, 89, 90, 91, 93, 97, 120, 124, 125, 126, 134, 138, 141, 145, 146, 147, 148, 149, 150, 153
liability, x, 22, 27, 51, 53, 54, 83, 91, 96, 97, 100, 103, 107, 108, 109, 110, 111, 112, 114, 115
licenses, 2, 88, 89, 99, 112, 143, 149, 150
life cycle, 64
likelihood, 51
limitation, 13, 49
listening, 8
litigation, vii, x, 1, 2, 3, 5, 6, 7, 8, 16, 17, 19, 24, 26, 28, 51, 53, 59, 67, 87, 98, 103, 107, 115, 124
living standards, 79
local government, xi, 61, 88, 89, 137, 139, 140, 141, 142, 146
location, 132
lower prices, 62

M

management, 79
mandates, 11, 42, 44, 46, 57, 114
manipulation, 65
manufacturing, xi, 7, 22, 26, 78, 82, 87, 89, 91, 93, 97, 120, 124, 125, 127, 131, 132, 134, 138, 140, 145, 146, 148, 151, 152
manufacturing companies, 89, 146
marginal social cost, 41
market, 4, 5, 6, 15, 25, 32, 41, 42, 50, 53, 54, 55, 56, 57, 58, 59, 60, 61, 62, 63, 64, 65, 66, 67, 68, 78, 80, 81, 82, 86, 87, 91, 97, 98, 102, 124, 127, 130, 131, 132, 139, 140, 149
market failure, 4
market share, 62, 65, 78
market structure, 41
marketing, viii, 6, 30, 39, 52, 53, 54, 55, 56, 57, 60, 63, 64, 65, 67, 68, 130, 140, 149
markets, viii, 4, 5, 8, 15, 25, 32, 39, 40, 59, 60, 130, 132, 139
meanings, 139
measures, 4, 91, 133, 152
media, ix, 107, 109, 117
Medicare, 75
medication, 43, 49, 50, 53
membership, 142
mergers, 66
mice, 44, 45
microelectronics, 145
Microsoft, 7, 31
microwaves, 139
military, 132
minority, 6
missions, 150, 151
mode, vii, 2, 3, 7, 16, 29, 46
models, 102
modernization, 129
money, 129, 146, 149, 151
monopoly, 4, 14, 40, 41, 42, 45
monopoly power, 40
motion, 18, 108
motives, 17
movement, 128, 139
music, x, 108, 113, 114
music industry, 114

N

National Aeronautics and Space Administration, 88, 143
National Bureau of Standards, 144
National Institutes of Health, 61, 73
National Research Council, 30, 37
needs, vii, 1, 5, 9, 82, 89, 121, 122, 124, 132, 141, 142, 147, 152
network, x, 107, 108, 110, 112, 113
networking, 112, 142
next generation, 115
Nobel Prize, 139
novelty, 28, 42

O

obligation, 15, 17, 27
observations, 26
organization, xi, 41, 79, 124, 137, 139, 142, 143, 150
organizations, 42, 86, 89, 108, 121, 125, 127, 147, 149
orientation, 80
outline, 18
output, 8
ownership, vii, 1, 4, 10, 40, 41, 52, 53, 62, 66, 121, 132, 143, 144

P

partnership, 88, 128, 144
patents, vii, viii, 1, 2, 3, 4, 5, 6, 7, 8, 9, 10, 11, 14, 16, 20, 21, 22, 23, 24, 25, 26, 27, 29, 30, 32, 34, 39, 40, 42, 43, 44, 45, 48, 49, 50, 51, 52, 53, 55, 56, 57, 58, 59, 62, 63, 66, 67, 72, 80, 87, 88, 89, 97, 98, 100, 102, 104, 128, 139, 144, 149, 150
PCR, 102
peptides, 99, 103

permit, x, 4, 12, 15, 41, 42, 64, 65, 119, 144, 148, 151
personal computers, 112
perspective, 18
pharmaceuticals, 40, 58, 60, 61, 62, 63, 67, 68
pharmacokinetics, ix, 95, 99, 101, 105
pharmacology, ix, 95, 101
philosophers, 54
photonics, 145
planning, 80
plastics, 139
police, x, 4, 12, 108, 110, 111
policy levels, 80
political parties, 152
poor, 6, 10, 24
population, 60
portfolio, 67
power, 8, 18, 45, 48, 53, 110, 124, 130, 145
predicate, 100
preference, 91
prejudice, 13
preparation, 98
President Clinton, 80, 82, 145, 146, 148
price competition, 62
prices, 6, 54, 62, 64, 65
principle, 10, 11, 12, 14, 22, 27, 50, 66, 114
private sector, ix, x, xi, 40, 61, 77, 78, 79, 80, 81, 82, 83, 84, 86, 87, 88, 89, 92, 93, 119, 120, 121, 122, 123, 124, 127, 128, 129, 130, 132, 133, 134, 137, 138, 139, 140, 141, 142, 143, 145, 147, 149, 150, 151, 152
private sector investment, 40, 84, 124
probability, 51, 114
producers, 24, 65, 121
production, 4, 6, 45, 51, 87, 121, 124
production costs, 121
productivity, viii, 4, 41, 77, 79, 80, 82, 122, 132, 140
profits, 26, 41, 51, 61, 66, 67, 123, 143
program, 8, 78, 80, 83, 84, 85, 87, 88, 89, 90, 91, 108, 111, 112, 116, 117, 124, 125, 126, 127, 128, 131, 133, 134, 141, 143, 145, 146, 147, 148, 150, 152
programming, 113, 116
proliferation, 129
promote innovation, 86, 141, 148
property rights, 25
proposition, 111, 149
proteins, 98
public health, 24, 144
public interest, 24, 51
public investment, 65
public policy, 91, 97, 112
public sector, 40, 139, 140, 141, 149

Puerto Rico, 89, 147

Q

qualifications, 35, 48
quality improvement, 82
query, 112

R

race, 11, 67
range, 5, 12, 28, 43, 63, 79, 81, 127, 140, 145
rate of return, 41, 61, 79
reading, 114
reagents, 102
real estate, 25
reality, 42, 59, 101
reasoning, 22, 44, 57, 101, 106, 110
receptors, 98, 99
recognition, 5, 62, 88, 89, 122, 141, 150
recovery, 52
reduction, 16, 36, 41, 78
reforms, vii, 1, 2, 5, 7, 10, 15, 29, 68
regulation, 20, 21, 98, 103
regulations, 18, 21, 57, 67, 82, 101, 144
regulatory requirements, 63
rejection, 20, 44, 47
relationship, 5, 21, 60, 81, 115
relationships, 123
relevance, 19, 20
rent, 6, 8
replacement, 62
reproduction, 109
Republicans, 91
research funding, 61
resistance, 80
resolution, 57
resources, x, xi, 4, 8, 11, 19, 23, 25, 41, 42, 80, 81, 87, 89, 92, 119, 121, 122, 123, 129, 131, 137, 139, 141, 142
responsibility, xi, 46, 81, 88, 111, 124, 137, 142
retail, 61, 62
retinopathy, 99
returns, 42, 65, 79, 80, 150
revenue, 7, 62, 110, 111
rewards, 67, 140
rheumatoid arthritis, 99
rickets, 24
rights, vii, viii, 1, 2, 3, 4, 5, 6, 7, 8, 9, 11, 12, 13, 15, 17, 18, 19, 21, 23, 25, 27, 29, 30, 32, 39, 40, 42, 44, 47, 48, 49, 50, 54, 55, 56, 58, 63, 86, 89, 96,

100, 102, 103, 109, 115, 117, 127, 139, 144, 149, 150
risk, 32, 53, 82
Ronald Reagan, 55

S

safety, ix, 45, 53, 55, 56, 60, 71, 95, 97, 101, 104, 144
sales, 12, 20, 42, 57, 61, 62, 65, 66, 67
savings, 68
scarce resources, 91
search, x, 20, 45, 46, 65, 108, 110, 112, 113, 117
searches, 112
searching, 109, 111
security, 5, 123, 127, 140, 151
seed, 84, 124, 145
self, 3, 112
semiconductor, 9, 32, 43, 127, 130, 132
Senate, 74, 75, 83, 84, 85, 90, 91, 92, 93, 125, 126, 133, 134, 135, 138, 145, 146, 147, 148, 153
separation, 85
series, 21, 43, 79, 85, 122, 128, 146, 148
serum, 54
service provider, 8, 25, 114
services, x, 8, 25, 32, 46, 78, 79, 81, 82, 85, 91, 108, 114, 119, 121, 130, 132, 141, 143
sewage, 24
shape, 123
sharing, ix, x, 88, 107, 108, 110, 111, 112, 113, 114, 115, 116, 119, 144
side effects, 105
signals, 129
skills, x, 3, 87, 119, 122, 123, 128, 139
small firms, vii, 1, 5, 8, 9, 11, 32, 83, 125, 149
software, ix, 9, 10, 107, 108, 109, 110, 111, 112, 113, 114, 115, 116, 123, 125, 145
specific knowledge, 111
specificity, 99
speculation, 2
speed, 8, 122, 132
spillovers, 29
spin, 139
sports, 113
stability, 54
stages, 98, 131, 140
standard of living, 78, 121, 140
standards, 3, 9, 24, 112
statutes, 40, 49
steel, 128
stimulus, 62
stock, 41
storage, 145

strategies, 78
strength, 32, 43, 78
subpoena, 18
subsidy, 25
substitutes, viii, 8, 40, 61, 62, 64
summaries, 101
superiority, 81, 151
supervision, x, 108
suppliers, 57, 122
supply, 44, 54, 100, 101, 141
support services, 116
Supreme Court, ix, x, 3, 14, 44, 45, 50, 51, 57, 95, 96, 98, 100, 101, 102, 103, 104, 105, 107, 109, 111, 112, 113, 115
surprise, 21
Switzerland, 34
syndrome, 87, 139
systems, x, 13, 107, 111, 132, 145

T

tax credit, 60, 78, 80, 83, 86, 92, 120, 126, 127, 134, 135, 138, 151, 152, 153
tax deduction, 86, 126, 127
technical assistance, 128
technological advancement, vii, viii, x, 1, 4, 8, 41, 77, 78, 79, 81, 82, 92, 119, 120, 121, 122, 123, 124, 129, 144, 152
technological change, 32
technological progress, vii, ix, 1, 2, 42, 78, 79, 85, 121, 140
technology, vii, viii, x, xi, 2, 4, 5, 7, 8, 9, 16, 19, 22, 29, 32, 40, 42, 47, 77, 78, 79, 80, 81, 82, 83, 85, 86, 87, 88, 89, 91, 107, 112, 113, 119, 120, 121, 122, 123, 124, 125, 127, 128, 129, 130, 132, 133, 137, 138, 139, 140, 141, 142, 143, 144, 145, 146, 147, 149, 151, 152
technology transfer, vii, xi, 5, 9, 80, 88, 89, 120, 128, 132, 137, 138, 139, 140, 141, 142, 143, 144, 147, 149, 151, 152
television, 113, 116
test data, 55
theory, 42, 62, 79, 111
thinking, 92
time, viii, xi, 2, 3, 4, 5, 7, 9, 10, 14, 16, 17, 19, 20, 21, 25, 29, 30, 32, 41, 42, 43, 44, 46, 48, 52, 53, 54, 56, 57, 61, 62, 63, 64, 66, 68, 75, 86, 89, 97, 109, 111, 112, 113, 116, 120, 127, 130, 137, 140, 143, 148, 149, 150
time constraints, 66, 89, 150
timing, 10, 20
tissue, 45
toxicity, 99, 105

trade, 4, 8, 12, 19, 21, 22, 55, 68, 78, 79, 121, 132
trade-off, 55
trading, vii, 1, 2, 5, 7, 13, 23, 29
trading partners, vii, 1, 2, 5, 7, 13, 23, 29
traffic, 110, 112
training, 49, 86, 127, 150
traits, 6
transaction costs, 4
transactions, 2, 4, 6, 112
transition, 25, 48
transition period, 25, 48
translation, 19
transmission, 108
trend, 8, 82
trial, 27, 97, 99, 100, 101
tumor(s), 44, 45, 99
tumor growth, 99

U

U.S. economy, vii, viii, 2, 5, 29, 32, 79, 81, 132, 140
uncertainty, 21, 80, 103, 104, 112
United States, viii, ix, x, 2, 3, 8, 10, 11, 14, 15, 19, 23, 29, 30, 33, 34, 39, 43, 45, 47, 49, 50, 51, 52, 53, 54, 55, 57, 58, 60, 61, 71, 72, 78, 79, 80, 81, 86, 87, 95, 96, 98, 106, 119, 120, 124, 127, 128, 130, 132, 139, 144, 150
universities, vii, x, 1, 5, 8, 9, 11, 82, 83, 84, 85, 86, 88, 89, 93, 102, 119, 120, 121, 123, 124, 126, 127, 128, 134, 138, 141, 143, 144, 145, 149, 150, 151
Uruguay, 59, 149

Uruguay Round, 59, 149

V

validity, viii, 3, 5, 6, 12, 16, 24, 27, 28, 32, 39, 51, 52, 57, 58, 104
values, 32
vapor, 145
variables, 140
venture capital, 67
vessels, 99

W

wages, 83
war, 123
weapons, 151
websites, 108
welfare, 41, 82, 123, 140
well-being, 151
witnesses, 6
words, 24, 42, 139
work, vii, x, 1, 17, 27, 29, 40, 41, 42, 57, 74, 79, 82, 83, 86, 88, 90, 91, 100, 102, 103, 105, 119, 121, 122, 123, 126, 128, 129, 132, 140, 142, 143, 144, 149, 151
World Trade Organization, 72
World War I, 85, 123
wound healing, 98
writing, 52